PUBLIC HEALTH AND
PRIVATE WEALTH

PUBLIC HEALTH AND PRIVATE WEALTH

Stem Cells, Surrogates, and
Other Strategic Bodies

Edited by
Sarah Hodges
Mohan Rao

OXFORD
UNIVERSITY PRESS

OXFORD
UNIVERSITY PRESS

Oxford University Press is a department of the University of Oxford.
It furthers the University's objective of excellence in research, scholarship,
and education by publishing worldwide. Oxford is a registered trademark of
Oxford University Press in the UK and in certain other countries.

Published in India by
Oxford University Press
YMCA Library Building, 1 Jai Singh Road, New Delhi 110 001, India

© Oxford University Press 2016

First published by Oxford University Press India in 2016

ISBN-13: 978-0-19-946337-4
ISBN-10: 0-19-946337-9

Typeset in Bembo Std 10.5/13
by The Graphics Solution, New Delhi 110 092
Printed in India by Rakmo Press, New Delhi 110 020

To all the students who were part of this project
—Sarah Hodges

*To Imrana Qadeer and Betsy Hartmann, gentle mentors
and amazing friends*
—Mohan Rao

Abbreviations

ABLE	Association for Biotech Led Enterprises
AIIMS	All India Institute of Medical Sciences
ARI/ARTI	Annual Risk of Tuberculosis Infection
ART	assisted reproductive technology
BCIL	Biotech Consortium of India Limited
BIPP	Biotechnology Industry Partnership Programme
BMC	Bombay Municipal Corporation
BMRC	British Medical Research Council
BT	biotechnology
CII	Confederation of Indian Industry
CMC	Christian Medical College
CSIR	Council of Scientific and Industrial Research
CT	charitable trust
CTRI	Clinical Trial Registry of India
DAE	Department of Atomic Energy
DALY	Disability-Adjusted Life Year
DBT	Department of Biotechnology
DOTS	Directly Observed Treatment, Short course
DPT	Diphtheria, Pertussis, and Tetanus
DST	Department of Science and Technology
FICCI	Federation of Indian Chambers of Commerce and Industry
GAVI	Global Alliance for Vaccines and Immunisation
GBD	Global Burden of Disease
GOI	Government of India
HDI	Human Development Index
HSPH	Harvard School of Public Health

HSPR	Health Sector Priorities Review
ICAR	Indian Council of Agricultural Research
ICMR	Indian Council for Medical Research
IHB	International Health Board
IK	Indigenous Knowledge
INR	Indian Rupees
INSTAR	Indian Society of Third Party Assisted Reproduction
inStem	Institute for Stem Cell Biology and Regenerative Medicine
IPD	in-patient department
IPPF	International Planned Parenthood Federation
IPR	Intellectual Property Rights
IT	information technology
IUAT	International Union Against Tuberculosis
JHSPH	Johns Hopkins School of Public Health
KNVC	Royal Netherlands Tuberculosis Association
MCGM	Municipal Corporation of Greater Mumbai
MDR-TB	multidrug resistant TB
MGH	Municipal General Hospital
MoU	memorandum of understanding
NBTB	National Biotechnology Board
NCBS	National Centre for the Biological Sciences
NIEO	New International Economic Order
NRDC	National Research Development Corporation
NTI	National Tuberculosis Institute
NTP	National Tuberculosis Programme
OPD	out-patient department
PC	power company
PPP	public–private partnership
PSU	public sector undertaking
PUPFIP	Protection and Utilization of Public Funded Intellectual Property
R&D	research and development
RSS	Rashtriya Swayamsevak Sangh
SBIRI	Small Business Innovation Research Initiative
SRMTE	Society for Regenerative Medicine and Tissue Engineering
STI	Science, Technology and Innovation

TB	tuberculosis
TCC	Tuberculosis Chemotherapy Centre
TSRU	Tuberculosis Surveillance Research Unit
UGC	University Grants Commission
UHC	Universal Health Care
UK	United Kingdom
US	United States
WDR	World Development Report
WHO	World Health Organization

Acknowledgements

Thanks to the British Academy for an International Partnership Award to the University of Warwick, UK, and Jawaharlal Nehru University (JNU), New Delhi. This award supported the six meetings from which these essays grew. Thanks also to the University of Warwick for administering these funds. In particular, these meetings would not have been possible without the labours of Liese Perrin, Dave Duncan, Katie Klaassen, and Frank Gallacher.

In Delhi, thanks to one and all at JNU—a university that continues to inspire. In particular, thanks to the Centre for Social Medicine and Community Health, including Singh Sahib.

Mohan Rao would gratefully like to acknowledge all his colleagues and students at the Centre of Social Medicine and Community Health, JNU, who made the workshops possible and memorable.

The editors would like to thank the team at Oxford University Press for the moral and practical support they gave the project.

Sarah Hodges
Mohan Rao

Introduction

*Science, Technology, and Medicine in India:
The Problem of Poverty*

Sarah Hodges and Mohan Rao

As 'drain theory' in 1901 or 'garibi hatao' in 1971, the eradication of poverty was the predominant economic, political, and social paradigm through which India's late colonial, nationalist, and post-independence critics and leaders constructed their legitimacy. Whether as critics of India's poverty, or as architects of measures for its eradication, India's commentators called on a broad framework of 'science' both to diagnose and treat poverty. Looking back, we are today able to pose a number of awkward questions about the careers of science, technology, and medicine in mediating poverty in India. Not only does the problem of poverty appear unsolved today, in contrast to a century ago, poverty eradication as a goal in itself seems to have fallen off India's scientific agenda almost entirely. How do we account for this sidelining of poverty? Has there been a loss of faith in the ability of science to tackle poverty? Or, in light of India's recent economic successes, has the problem of poverty simply become inconvenient?

As significant pieces of new research, the chapters in this volume address two interlinked challenges. First, these chapters are written as initial attempts to problematize the concept of poverty as it is currently found—or obscured—within expert economic enquiry. Economistic models, particularly those of cost-benefit analyses, have regularly set the boundaries for imagination of schemes to address poverty and ameliorate the plight of the poor. This lead to a set of technocratic poverty-control methods that have often been seen to succeed on the sole basis that they deliver information that fits neatly into data sets. And, in turn, this econometric data stands as some sort of 'complete' knowledge regarding both the dynamics of poverty and the plight of the poor. In so doing, letting economists rule the kingdom of poverty has both obscured our view and limited our abilities to challenge this work and to frame different questions. As Breman has written:

> The study of poverty is nowadays considered to be the business of economists. Or at least those among them engaged full-time in the practice of econometrics and statistics. They do not easily accept that the classical literature on the subject of human deprivation has been produced by political economists, sociologists, anthropologists and historians.[1]

In other words, rather than needing or leading from a 'definition' of poverty—particularly given that definitions of poverty are not now, and likely have never been, empirically self-evident or politically neutral—the chapters that follow are guided by a different set of questions: What question is poverty the answer to? In other words, what use is the concept? What 'work' does poverty do? Within expert discourse, whom or what does the concept empower? What does it obscure?

Second, the chapters in this volume are written in response to the somewhat surprising absence of 'poverty' as a central object of analysis despite the scholarly dynamism that characterizes much contemporary study into the careers of science, technology, and medicine in modern India. In other words, we now have a growing body of scholarship on India that considers science, technology, and medicine alongside and as part of colonialism, nationalism, Nehruvian India, and post-liberalization India. Albeit in very different ways, in each of these periods and projects the problem of poverty—and the poor—was a key concern for policy makers and others tasked both with organizing science and ruling India.

Third, what do we make of India after liberalization? There is much evidence that the health sector reforms that have been undertaken

---

[1] Jan Breman, 'At the Bottom on the Urban Economy. Review—Ananya Roy, City Requiem, Calcutta: Gender and the Politics of Poverty', *Economic and Political Weekly* 38, no. 39 (27 September 2003): 4138.

had had dolorous outcomes, that the insurance-led model had indeed contributed to increasing out of pocket health expenditures, and impoverishment, that it had set out ostensibly to reduce[2] and that it had contributed both to inappropriate medical care while transferring public resources to the corporate sector in medical care.[3] The moral hazard problem of this model of medical care was also highlighted by studies indicating unnecessary surgeries, hysterectomies in particular and unnecessary investigations. For example, 16,765 women in Bihar and 1,800 women in Chhattisgarh had had their uteri removed unnecessarily by private doctors in order to claim insurance.[4] That there was a need to strengthen the public health system and move away from this model of health care delivery had been acknowledged by the report of the Planning Commission's High Level Expert Group.[5] That this model of health care expansion is at the behest of institutions that represent insurance and the corporate sector in medical care is evident from the interest that the Confederation of Indian Industry (CII), which represents these interests, is able to influence health policy.[6]

The rush to commercialize India's health care system received a new boost with the 2014 election of the Bharatiya Janata Party (BJP) government, which is committed to growth. The Draft National Health Policy 2015 notes with approbation that the private health care industry is valued at US$40 billion and is projected to grow to US$280 billion by 2010 as per market sources. The current growth rate of this

[2] Sakthi Selvaraj and Anup K. Karan, 'Why Publicly-Financed Health Insurance Schemes Are Ineffective in Providing Financial Risk Protection', *Economic and Political Weekly* 47, no. 11 (2012): 60–8.

[3] Rajan Shukla, Veena Shatrugna, and R. Srivatsan, 'Aarogyasri Healthcare Model: Advantage Private Sector', *Economic and Political Weekly* 46, no. 49 (2011): 38–42.

[4] 'A "Twisted" Health Scheme: State Health Insurance Schemes Are Being Abused to Profit Private Care Providers', editorial, *Economic and Political Weekly* 47, no. 38 (2012): 8.

[5] Government of India, Planning Commission, *High Level Expert Group Report on Universal Health Coverage for India* (New Delhi, November 2011).

[6] Vidya Krishnan, 'Private Practice: How Naresh Trehan Became One of the India's Most Influential Doctor-Businessmen', *The Caravan*, 1 February 2015, http://www.caravanmagazine.in/reportage/naresh-trehan-medanta-private-practice.

perennially and most rapidly growing area of the economy, the health care industry, at 14 per cent is projected to be 21 per cent in the next decade. Indeed in one year alone, 2012–13, as per market sources, the private health care industry attracted over two billion dollars of FDI, much of it venture capital. For International Finance Corporation, the section of the World Bank investing in the private sector, the Indian private health care industry is the second highest destination for its global investments in health.[7]

It is thus not surprising that the recent budget sees no increase in India's abysmal public expenditure on health. Indeed, given inflation, it represents a decrease in health care allocation, already one of the lowest in the world.

Chronologically speaking, the volume's chapters attempt to redress these absences through addressing three broad problematics that under-pinned the relationship between science and poverty as it unfolded across the twentieth century. Part 1 explores the relationship between science, technology, and medicine, on the one hand, and the quest for 'improvement' in late colonial and early independent India, on the other hand. Part 2 explores the varied careers of the hospital in independent India. It asks to what extent has everyday health care delivered on its promise to improve the everyday lives of ordinary Indians. Part 3 takes up the question of the history of official science policy—particularly the career of biotechnology—to ask whether, by the close of the twentieth century and the opening of the twenty-first, India's poor came to figure most strategically in science policy's new tryst with private industry as experimental material. Finally, at the risk that this endeavour is mistaken for a critique of the West, the volume closes with a critical review of the so-called indigenous turn in science.

PART 1   SCIENCE, POVERTY AND THE QUEST FOR 'IMPROVEMENT'

Taken together, the chapters that form the opening section of the volume—'The Quest for "Improvement"'—point to how the optic of poverty was at the forefront of international scientific planning in

---

[7] Government of India, Ministry of Health and Family Welfare, *Draft National Health Policy 2015* (New Delhi, December 2014).

India, from colonialism to international development in Nehruvian India. This section of the volume opens with a chapter by David Arnold: 'Colonial Poverty: Nutrition, Disease, and the Problem of the Poor'. Arnold's chapter denaturalizes the connections between poverty and public health through a close examination of the career of poverty in official policy in India over the course of a century—from the 1850s to the 1950s. In so doing, he shows that poverty was certainly a recurring theme in India's colonial governance and emergent public health regime. Yet this material also reveals a perhaps unexpected point. Although colonial medical professionals recognized widespread poverty among the Indian population, they did not also see poverty per se as a fundamental cause of ill-health and mortality. Nor, Arnold argues, did they feel empowered to offer much by way of practical measures for its amelioration. Poverty was essentially seen as a background factor in the causation and incidence of disease or as the consequence of disease rather than one of its underlying causes. A new recognition of the importance of poverty to public health really emerged in India only after 1918, especially through the study of nutrition. It was in this area—that could not avoid questions of basic access to food—where there developed a clear set of poverty-related public health methods and objectives. However, Arnold closes with the observation that, as India entered a new era of state planning mid-century, nutritional analyses of—and nutritional solutions to—the corporeal effects of poverty were overshadowed by strategies that disaggregated 'the problem of poverty' into a set of highly technologized solutions. Arnold's chapter matters because it intervenes in conventional wisdom about the career of the anti-politics of poverty—characterized by a reticence to frame or address poverty as a political rather than a narrowly technological problem. That is, most studies see the rise of an anti-politics of poverty as part and parcel of the consolidation of developmentalist regimes after the middle of the twentieth century. In contrast, Arnold points to a far longer history of an anti-politics of poverty as it emerged in India; a longer history that is as much part of Britain's own history of poverty 'at home' as in the colonies.

Nevertheless, it is certainly the case that technocratic solutions to poverty found full flower in independent India's experience with development as a cure to poverty. As the other two chapters in this section show through their respective analyses of the careers of tuberculosis and population control, technologies both produced and were produced by a narrowing of the possible realms for intervention into poverty. Lakshmi Kutty, in her chapter 'Tubercular Optics: Health, Techno-science, and the Obfuscation of Poverty', notes that tuberculosis (henceforth TB) has long been acknowledged as the quintessential disease of poverty. She narrates how, in the middle of the twentieth century, drug treatments emerged that could control TB in the short-term. These were seized on by both India's national government and international agencies that had long seen TB as one of the biggest obstacles to achieving an efficiently functioning and productive economy. By the 1980s, however, there was a policy reconsideration. The approaches adopted had not yielded dividends. Not only had the problem of TB not been brought under control, there was a resurgence of TB. Dominant policy understandings of TB moved away from political questions of poverty, deprivation, or mounting inequalities. In their place, health policy workers attempted to work round poverty by casting the enduring TB rates as an effect of inadequate implementation of techno-scientific solutions. Earlier questions of a political nature were systematically excluded.

The overview of twentieth-century TB interventions matters because it demonstrates the sea change in the strategic understanding of the link between poverty and ill-health. Kutty's account of the changing fortunes of TB control strategy in India adds a needed update to earlier McKeownite debates about the relationship between nutrition and health across time. Briefly put, the McKeown thesis was that large scale improvements in population health (particularly in Europe) predated modern medical interventions or cures. Rather than taking up this debate in the aggregate, Kutty shows how, even if McKeown was wrong and medical breakthroughs were behind improvements in health over the course of the nineteenth century in Europe, this argument could not be used to support TB programmes in India. In the absence of basic levels of nutrition, the technological solution to TB (taking medicine) is simply ineffective. Malnourished 'poor bodies' do not respond to drug therapies. By systematically excluding a broader political approach to poverty eradication from programmes to control TB, the specifics of TB treatment itself could not improve health but in fact gravely undermined the possibilities of such intervention to improve the lives of the poor, let alone heal their bodies.

Like TB, in debates over the reasons behind changes in population health over time, population control emerged as a key technology in the global 'war on poverty' as it emerged in the 1960s and 1970s. Population—or rather, 'overpopulation'—was one of the most central and enduring explanations for the endurance of poverty in India and elsewhere across the globe. Alongside the Green Revolution, population control at mid-century was initially discussed as a measure to curtail the growth of poor populations in order to create a new world, post Green Revolution, in which it was possible to feed everyone. However, population control programmes quickly grew into a project in and of itself. Rebecca Williams's chapter highlights how international studies of India's fertility were central in the formation of global population policy. During the 1970s, health inequalities became a key issue for international health experts.

In the chapter "'Surveillance for Equity"? Poverty, Inequality, and the Anti-politics of Family Planning', Rebecca Williams turns to the system of surveillance promoted as a means to rectify health inequalities within one of the earliest and most influential population studies—the Khanna study. This well-known 'applied research' population control experiment was run in the Ludhiana District of Punjab during the 1950s by researchers from Harvard. The Khanna study took a system of continuous population monitoring, derived from communicable disease study, and applied it to population control. By tracing the adaptation of this mode of surveillance from the Khanna study to health 'equity', this chapter shows that 'surveillance for equity' was not a new approach to health care, but an existing set of technical interventions repackaged in the language of social justice. In particular, this chapter argues that by focusing narrowly on health inputs and outputs, the model of health equity has systematically erased the political and structural causes of poverty, and replaced them with technical causes, requiring technical solutions. In short, 'surveillance for equity' isolated the bodies of the poor from their social, economic, and political contexts, and transformed them into objects of medical knowledge and management.

PART 2   INDIA'S HOSPITALS: FOR WHOM?

If the chapters in Part 1 address the relationship between policy planning and micro-practice of that characterized 'developmentalist' projects such as disease eradication and population control, the three chapters of Part 2 constitute an in-depth look at how hospitals—and clinical health care more generally—have mediated the everyday relationships among poverty, medicine, and state policy in independent India. Hospitals matter because the clinical spaces are where a great majority of Indians experience the so-called fruits of medical science. What is less obvious is how hospitals matter to poverty. Taken together, these chapters make for uncomfortable reading. They point to how Indians' everyday experience of the 'fruits of medical science' have become ever confined to those who can pay. The state seems to have abandoned a health care policy explicitly aimed to provide medical service for all Indians, including the poor. By looking at the careers of everyday health care policies, these chapters ask if, by the closing decades of the twentieth century, the effects of health policy served less to enable the health care sector to serve the health of India's people, but instead to underwrite the health of India's economy.

This section opens with Ramila Bisht and Altaf Virani's chapter 'Globalization and the Health of a Megacity: The Case of Mumbai'. This chapter reviews the health care landscape of Mumbai over the last few eventful decades through a close examination of a public–private partnership (PPP) in the city that converted a dysfunctional municipal maternity home into a multi-specialty peripheral hospital, ostensibly to benefit the slum-dwellers in the catchment area. In providing an overview of Mumbai's health care system, this study evaluates the effects of a two-track (public and private) health care on the health of the city's residents, both rich and poor. In particular, the chapter tracks changes in the city's health care infrastructure over the last two decades, giving particular attention to the formulation and operationalization of new policy initiatives' push towards privatization of the public health system. Bisht and Virani's material shows that these changes have substantially undermined the provision of health care for the poor. In fact, rather than extending high quality health care to underserved populations, Bisht and Virani find that the first public health care PPP of its kind in Mumbai created greater obstacles for the poor in accessing health care at all.

This chapter matters to our understanding of the everyday impact that new policy directions have on the lives of the poor. From the closing decades of the twentieth century, support for private health care across the globe grew with the initiation of PPPs. In the international

policy arena, global PPPs emerged as the new mantra in a partnership between an underfunded WHO and new international NGOs,
particular NGOs aimed at addressing global public health concerns of
HIV/AIDs, TB, and malaria.[8] Within the many state funding crises of
the 1990s—in India and across the globe—PPPs were promoted as a
way of revitalizing public health infrastructure. Although many PPPs
exist across India, and much anecdotal evidence exists, there have been
few empirical studies. Hence this chapter not only serves to question
received policy wisdom, but also, through its mixed method analysis,
suggests methodological ways forward in our understanding of the
everyday connections between policy and practice.

In 'Commercialization and the Poverty of Public Health Services
in India', Rama Baru traces the shifts in the relationship between the
public and private sectors in medical care. Her chapter is so significant
because she points out that rather than starting as a state project and
becoming a private enterprise, the line between private and public
health care delivery in independent India has often been blurred.
Attending to the critical distinction between 'privatization' and 'commercialization', Baru argues that even if public sector still has a role
to play in health care provision, the public and non-profit sector have
become 'commercialized' given how market principles have perme-
ated the organization of the public sector itself.

The significance of Baru's chapter also lies in how it situates the
health policies of post-liberalization India within a longer history. Her
work shows that the increased role of commercialization in informing everyday health care in fact is not completely new. Instead, Baru's
account of commercialization shows how the Indian health care scene
has long been a mixed economy. Baru's account also shows how, after
some expansion in the public health care sector in the 1980s, the main
driver of growth in health care capacity in India has been in the private
sector. Hence post-liberalization expansion of commercialization is in
part a continuation of long-standing relationships among health policy,
the Indian state, and the private sector. Yet more recent changes—such
as the late twentieth century innovations in high-tech medical diagnostics intensified earlier existing relationships of commercialization

---

[8] Von Eduardo Missoni, 'A Long Way Back towards Alma Ata', *Bulletin von
Medicus Mundi Schweiz* 111 (February 2009).

and exacerbated the disparities between the services available in the private and public health care sectors.

As in Bisht and Virani's analysis of PPPs in health care, Sarah Hodges critically examines another new—and largely celebrated—arrival onto India's health care scene: the corporate hospital. In '"It All Changed after Apollo" and Other Corporate Hospital Myths', Hodges uses material gathered from interviews with 20 prominent physicians in Chennai as well as materials gathered from newspapers and official and unofficial health-related documentation to examine a touchstone in India's recent health care history: that of the 1983 launch and subsequent three-decade career of Apollo Hospitals in Chennai. She argues that one of the singular successes of Apollo Hospitals (and its founder and Chairman, Prathap Reddy) has been in image management. In order to do this, Hodges explores the myths that surround Apollo and Reddy, and explore the corollary phenomenon of myth-making. In particular, she investigates not only the myth-making activities of Apollo and Reddy, but also how these stories are regularly reproduced within a wider (and often, ironically, a critically engaged) community of medical professionals across the region. To anticipate the argument somewhat, the Apollo 'success story', while central to today's dominant narrative of the economic growth produced by India's private health care sector, is based on assumptions and assertion. These assumptions and assertions begin to crumble under even the most basic historical scrutiny. In short, in its three decades, Apollo's greatest success may perhaps be its story.

Hodges' account matters to our understanding of broader trends within health care in India as they emerged in the closing decades of the twentieth century, because it is these trends that inform common-sense policy thinking today. The chapters of both Hodges and Bisht and Virani challenge the confidence placed in so-called corporate hospitals to deliver improved care to an entire population, not just to those who can afford to pay the high prices charged. Further, these private institutions are not organized around the principles of universal care. Instead, corporate hospitals have grown up as niche institutions that specialize in highly technologized and specific procedures for which middle class and rich Indians form a particularly profitable market (health of the heart, kidneys, and other chronic conditions). While expertise and outcome rates may be excellent, corporate hospitals have no track record in being able to provide effective treatment for more complex and less remunerative conditions for which poor patients more commonly seek treatment (gastrointestinal and respiratory illnesses among others). In light of contemporary debates unfolding around state insurance schemes, these chapters suggest that corporate health is better positioned to create profits rather than to provide for the health of the nation.

PART 3 NATIONAL TECHNO–SCIENCE AND PROMISING BODIES

Whereas the chapters in Part 1 explored how government policies addressed the compact between science and society through large scale studies and development programmes aimed at the poor, the chapters in Part 2 suggested how, despite the substantial growth of health care services after the 1970s, this excluded the poor from the expanding health care sector and the medical benefits it brought. On the heels of these changes, the chapters in Part 3 point to a recent and even more troubling trend for biomedical science in India, particularly in the field of biotechnology. Rather than new biomedical technologies serving the poor, it appears that the poor have emerged as one of India's biotechnology resources. The poor now serve medicine.

Did this inversion of the earlier relationships among science, service, and the poor happen out of the blue? Since when did commerce trump society in science planning? The answer to these questions is far from clear. For example, throughout the twentieth century, India's science policy—as elsewhere across the globe—was formulated with an eye to its potential practical applications, including industrial applications. One early example of this was the Council of Scientific and Industrial Research (CSIR), created in 1942 as part of a late colonial agenda for building infrastructure to support scientific research in India. However, despite such initiatives like the CSIR, the links between state research and industry tended to be episodic. Additionally, over the course of the second half of the twentieth century in India, and particularly by the 1970s and 1980s, industrial policy itself ceased to be a matter simply of national planning—either rhetorically or practically.

It is as part of this historical juncture that the example of biotechnology in India becomes instructive. By the 1980s, Government of India initiatives—such as the creation of the Department of Biotechnology in 1986—signalled a new role for scientific research in the relationship between state policy and private industry. As the chapters in Part 3 show, biotechnology has been a key site where state research policy has been crafted to enhance economic growth and national prestige. By the close of the twentieth century, rather than science functioning as a national enterprise (as it was seen to do in Nehruvian India), science is now driven largely by the agendas of private enterprise. As one observer summed up, 'science is no longer a part of [national] culture, it is now a part of [national] commerce'. One place this can be seen is in the national budget allocation for scientific research. During the four decades following independence in 1947, the Indian government's expenditure on science made real, year on year, increases. From the 1990s, however, this trend was reversed. Instead, the Government of India pursued collaborations with industry and expected that industry funds would underwrite national scientific research. What is less clear is what the knock-on effects of this policy have been.

In light of this, the chapters in Part 3 explore the varied careers of biotechnological research and commerce in India. In 'Biotechnology in India: Catalyst for a Knowledge Era?' Priya Ranjan places the development of India's medical biotechnology industry within the logic of global neo-liberal changes that shaped the very emergence of the biotech industry in the United States of America. Within this, those calling for the expansion of state support for biotech in India made an impassioned plea for a second chance for India to reap the benefits that accrue to nations that succeed in the linked pursuits of science, technology, and industry. These supporters argued that while India may have lost out in the nineteenth-century race for industrialization due to the constraints placed on the economy by British colonialism, biotech is the chance for India to come storming back. In other words, government officials claimed that, through the robust development of biotechnology research capacity, India would shake off its underdeveloped past and 'leapfrog' into a brave new high-tech future. As Priya Ranjan shows, supporters of expanding state support for biotech claimed that India could support biotech and become an international player by leveraging its wealth in human resources. Not only did India boast tech-savvy English-speaking scientific expertise, they made calls to exploit India's heterogeneous gene pool, the large number of diseases in the country, as well as the drug-naïve population that the poor 'gifted' the research sector.

In 'Stem Cell Research and Experimentation in India: Leveraging Hope for Global Prominence', Rohini Kandhari expands upon the broad context that Priya Ranjan's chapter provides. Kandhari examines stem cell research and commerce. She makes a number of startling discoveries. The first is the mismatch between the development of research and the development of commerce in India. Despite more than twenty years of state investment in India, basic research into stem cells remains at a very preliminary stage. In contrast, venture capital investments in stem cell therapeutics has surged ahead as they have exploited India's unregulated medical scene. Together, these factors have contributed to an extraordinary availability of doctors and hospitals offering speculative stem cell treatments. Kandhari examines how patients with diseases such as spinal cord injuries, cerebral palsy, Asperger's Syndrome, Parkinsonism, and so on are being offered stem cell therapy in a number of cities in India. Additionally, she explores how a public policy-led growth of medical tourism has also facilitated the stem cell industry. Patients from abroad—the so-called therapeutic refugees—can come to India because such therapies are not permitted in their own countries. In closing, Kandhari asks: how ethical is it for India to be investing in such technologies, when it is unwilling to make investments in the health care needs of the majority of the population?

Following on from Kandhari's discussion of biotechnological science in terms of stem cell research and commercial therapeutics, Mohan Rao turns to another manifestation of biotech business in India: commercial surrogacy. In his wide-ranging chapter, 'The Globalization of Reproduction in India: From Population Control to Surrogacy', Rao traces the making and re-making of the fertility of India's poor over the past few centuries: from the problem of so-called overpopulation to the 'solution' of surrogacy. Within a political economy of health approach, this chapter first describes how nineteenth century overpopulation theories dovetailed with India's surplus being sucked into Britain's empire. It goes on to show how nineteenth and twentieth-century eugenic arguments became intertwined with those of the idea of a surplus population. Building on this development, Rao shows how the resulting family planning goals after independence fundamentally skewed India's health care services. It concludes by setting the emergence of India's high-tech medicine in the context of its pursuit of neo-liberal economics over the past three decades. India has emerged as a major hub in what is today a global surrogacy industry. This chapter locates the emergence of surrogacy as the solution to the so-called population problem. In short, Rao argues that commercial surrogacy 'recycles' the historic problem of the poor into contemporary resources for the rich.

AFTERWORD: WHEN WAS KNOWLEDGE INDIGENOUS?

In the afterword, 'Mainstreaming Indigenous Knowledge: Genealogy of a Meta-concept', philosopher and historian Dhruv Raina examines the dual career of the category 'indigenous knowledge' within development practice and science studies. It shows how the interest in indigenous knowledge emerged in tandem with the politicization of indigenous groups and indigenous-rights movements, wherein the indigenous peoples began to demand the right to be heard in development-related decisions concerning themselves. As a result, 'indigenous knowledge' came to be lauded as an alternative collective wisdom relevant to a variety of matters at a time when existing norms, values, and laws have been increasingly called into question and in cases where development planning has not delivered on promised results. In other words, indigenous knowledge comes to form part of the project of sustainable development. Thus the possible mainstreaming of indigenous knowledge and indigenous science has begun in earnest as the recognition in the scientific community dawns that the problems of poverty and sustainability are far too complex and risky to be handled by what is received as big science today.

These chapters attempt to advance a different set of questions to guide the study of science and poverty in modern India. They insist that the commodification of science (particularly in health and medicine, the subject of many of the chapters) is both fundamentally about economies of bodies, yet irreducible to conventional economic frameworks. The subject—and the study—of poverty demands that we transcend traditional disciplinary boundaries, and combine the methodologies of anthropologists, sociologists, health economists, science studies scholars, and historians. In sum, by pursuing questions of who wins—and who loses—in India's scientific histories and economies, the chapters in this volume cut across studies of development, public health, and demography to explore how science has informed the everyday lives and livelihoods of the rural and urban poor from the opening of the twentieth century up till the present day.

Part 1

The Quest for 'Improvement'

1 Colonial Poverty

Nutrition, Disease, and the Problem of the Poor*

David Arnold

It is almost commonplace today for poverty to be identified with malnutrition, high levels of infant and maternal mortality, and the incidence of diseases like cholera, typhoid, and malaria. A correlation between poverty and the so-called tropical diseases has become almost axiomatic.[1] But this has not always been the case and it is worthwhile to consider when and how the concept of poverty began to impact on the medical and sanitary understanding in India and how poverty began to figure in emerging public health discourse and practice. An obvious point of reference for such an enquiry is Michel Foucault's discussion of governmentality, but this discussion also draws inspiration from Christopher Hamlin's account of social justice and the medicalization of poverty in nineteenth-century Britain and Giovanna

* A different version of this chapter was published in *Historical Research* 85 (2012): 488–504, under the title 'The Medicalization of Poverty in Colonial India'. I am indebted to Sarah Hodges and Mohan Rao for the opportunity to revise and extend some of the arguments made in that article.

[1] Göran Djurfeldt and Staffan Lindberg, *Pills Against Poverty: A Study of the Introduction of Western Medicine in a Tamil Village* (London: Curzon Press, 1975).

Procacci's insightful analysis of the 'government of poverty'.[2] Although her immediate concern is with the social critique of political economy, Procacci's reflections on 'the government of the poor' raise a number of pertinent questions about the relationship between poverty and power that are relevant to health policy and practice in colonial India. Of what use, she asks, was the concept of poverty? Who or what did it empower? And to what was it the antithesis? In British India, poverty has widely been understood in terms of those it destroyed and disempowered, especially in times of rampant famine, but it has seldom been addressed, outside political economy, as a significant concept and one that might have an explicit bearing on how public health issues were framed and articulated. Historically, too, among colonial writers, while exceptions can certainly be found, the very prevalence of poverty and its presumed naturalness in India tended to favour its neglect as a specific subject of enquiry, while for nationalist and Marxist writers poverty primarily served as an indication of colonial exploitation and indifference to Indian well-being. There is much, therefore, to be gained, analytically and empirically, by foregrounding poverty as a primary issue of public health and a cardinal feature of colonial medical governmentality.

ENDEMIC POVERTY

References to poverty occur periodically in early colonial accounts of disease, but generally as only one of several background factors or 'predisposing' causes. Thus, in the cholera epidemics that repeatedly swept India from 1817 onwards, fatigue, debility, exposure, and want were identified as contributing to mortality among the poor without thereby being assigned a determining role.[3] Occasionally, poverty was

---

[2] Michel Foucault, 'Governmentality', in *The Foucault Effect: Studies in Governmentality*, ed. Graham Burchell, Colin Gordon, and Peter Miller (Chicago: Chicago University Press, 1991), 87–104; Giovanna Procacci, 'Social Economy and the Government of Poverty', in *Foucault Effect*, 151–68; Christopher Hamlin, *Public Health and Social Justice in the Age of Chadwick: Britain, 1800–1854* (Cambridge: Cambridge University Press, 1998).

[3] Frederick Corbyn, *A Treatise on the Epidemic Cholera as It Has Prevailed in India* (Calcutta: Thacker, 1832); James Jameson, *Report on the Epidemick Cholera Morbus* (Calcutta: Government Gazette Press, 1820).

foregrounded. In 1856, James Ranald Martin, a leading proponent of public health in India, remarked: 'Let want, filth, crowding, and misery be removed, and epidemics will have lost their chief power.'[4] But, in the main, it was 'the poor' (as victims) rather than 'poverty' (as cause or concept) that informed epidemiological discourse.

For most of the nineteenth century medical understanding of disease and its impact on public health followed one of three routes. First, with some contagious diseases, like smallpox, questions of poverty were seldom raised: their mode of transmission and the nature of their victims tended to preclude analysis in terms of wealth and poverty. Second, in the case of cholera, malaria, and other fevers, it was again not poverty per se but location and climate that were seen as the prime generators of disease, especially given the supposedly pathogenic nature of 'tropical' environments.[5] In a revivified Hippocratic tradition, poverty counted for less than the deleterious effects of airs, waters, and places. Third, this environmentalist understanding of disease coexisted with an 'othering' of Indian society—in the belief that Indians' customs, habits, social practices, and beliefs helped foster or disseminate disease. This is evident from Martin's claim that caste was 'of itself an enormous injury to public health', because it was 'prejudicial to public happiness'.[6] Modes of dress, housing, diet, marriage customs, and religion all informed this reading of India's cultural pathology. Hence, in discussions of epidemic cholera, prominence was given over many decades to Hindu religious fairs and bathing festivals as foci from which disease was repeatedly disseminated. That pilgrims might be poor mattered less than their seemingly irresponsible and unsanitary behaviour.[7] It was argued, too, that poverty in India was distinct from

[4] James Ranald Martin, *The Influence of Tropical Climates on European Constitutions* (London: John Churchill, 1856), 344.

[5] David Arnold, 'India's Place in the Tropical World, 1770–1930', *Journal of Imperial and Commonwealth History* 26, no. 1 (1998): 1–21.

[6] James Ranald Martin, *Notes on the Medical Topography of Calcutta* (Calcutta: G.H. Huttmann, 1837), 49.

[7] David Arnold, 'Cholera and Colonialism in British India', *Past and Present*, 113 (1986): 138–42; Mark Harrison, 'A Question of Locality: The Identity of Cholera in British India, 1860–1890', in *Warm Climates and Western Medicine: The Emergence of Tropical Medicine, 1500–1900*, ed. David Arnold (Amsterdam: Rodopi, 1996), 133–59.

that found in the West: Indians could survive in conditions that would have been intolerable or unsustainable elsewhere, with minimal levels of food, clothing, and shelter. They were less likely, in consequence, to fall ill as a result of levels of poverty and deprivation to which they were already habituated. This was one of the ways in which poverty came to be naturalized rather than exceptionalized in India. As Charles Grant put it: 'The tropical climate minimizes the need for food and artificial warmth, and so simplifies the mere act of living.'[8] Poverty was not ignored, but it was subsumed within more dominant ways of thinking about Indian health and the supposedly primitive nature of India's society and economy.

In colonial India, where religion, caste, and community were afforded primary significance, the category of 'the poor' was more elusive than in Victorian Britain and the 'discovery of the poor' advanced on different lines. India had no poor law, and thus no obvious mechanism by which the poor could be categorized, institutionalized, and medicalized, though, more by default than design, hospitals, dispensaries, and jails might fill this lacuna. In Britain the term 'pauper' was frequently used not just to describe want and destitution but also to isolate individuals who, through dissolute habits, moral weakness, and reluctance to work, were perceived as being the authors of their own misery and squalor, and, potentially at least, a danger to others.[9] The poor as a social category were not afforded such prominence in India. Even the word 'pauper', though certainly used, was employed far more sparingly in India.[10] If in Europe 'pauperism' implied the threatening

---

[8] Charles Grant, 'On the Poverty of India in the Nineteenth Century', Willoughby Collection, Mss Eur. E 308/51, India Office Records [hereafter IOR], British Library, London (with thanks to Shruti Kapila for this reference). Cf. Frederick Henvey, *A Narrative of the Drought and Famine which Prevailed in the North-West Provinces during the Years 1868, 1869, and Beginning of 1870* (Allahabad: Government Press, North-West Provinces, 1871), 2: 'the necessities of life in a tropical climate are few and simple'.

[9] On poverty and definitions of the poor, see Gertrude Himmelfarb, *The Idea of Poverty: England in the Early Industrial Age* (London: Faber and Faber, 1984).

[10] For one example of the extended use of the term 'pauper', see T.S. Weir, 'Annual Report of the Health Officer', in *Annual Report of the Municipal Commissioner of Bombay 1877* (Bombay: Times of India Steam Press, 1878), 21.

and unruly behaviour of the poor, in India the poor, who in practice included a large proportion of the labouring, artisan, and cultivating classes, were less likely to be identified as an immediate source of social unrest and political danger.

India might lack a poor law, but it did have a series of devastating famines that, over the course of the nineteenth century, claimed many millions of lives. Famine rendered poverty urgent, clamorous, and visible, but in the minds of many British commentators it was connected to the vagaries of the Indian climate and the fragility of its agriculture. Like the jails, to which many of the hunger-struck flocked, famine relief camps provided privileged sites for medical observation of the pathology of the poor. Some of the empirical data collected informed pioneering medical tracts like Alexander Porter's *Diseases of the Madras Famine of 1877–78*. But the incidence of epidemics concurrent with famine tended to obscure the significance of poverty as a prime cause of mortality. For instance, a report written in the late 1870s by Henry Vandyke Carter and T.G. Hewlett of the Indian Medical Service on the recent outbreak of relapsing fever in Bombay attributed the disease to a 'specific poison' which had afflicted the city's poor and affluent alike, but it made no direct reference to the famine and destitution then sweeping the province. This drew the ire of Bombay's municipal Health Officer, T.S. Weir, who declared that the absence of any mention of the famine was an 'astonishing omission', as was the absence of any reference to high prices and food shortages in the city. The authors of the report had, he declared, avoided

> all allusion to the desperate straits to which the poor in this city were reduced…the great causes that had disturbed the public health did not excite even a passing remark. Whether they were wilfully ignored, or unconsciously passed over, the fact remains the same—they were left unnoticed.[11]

Weir's remarks are one indication among many (as will be seen more fully in the following section) of the way in which poverty attracted greater attention from the municipal health authorities, especially in metropolitan Bombay, than it did from the state-run medical service as a whole or in reports which sought to address epidemic causation

---

[11] Weir, 'Annual Report of the Health Officer', 47.

and mortality at large rather than local patterns of urban sickness and deaths. But, beyond Bombay, the argument for poverty as a major factor in explaining epidemic mortality tended to be marginalized, even where it was not directly contested.[12] This became especially so once, for example, relapsing fever had been identified as a tick-borne infection. It was declared in the early 1920s to be a disease not confined to 'the poorest and most ill-housed section of the population', but 'one to which practically all the inhabitants are liable, irrespective of their caste or habitat'.[13]

Public health tended to follow, not dictate, government policy. The government's view of famine was grounded in classical political economy and the principles of laissez-faire: even the famine poor were required to work for their subsistence, unless they were so debilitated as to render this physically impossible. Viewing famine primarily as an economic crisis, the state was concerned to minimize relief expenditure, prevent 'dependency' among the poor, rebuild the agrarian economy as quickly as possible and create an infrastructure that would render future famines, if not less likely, then at least less deadly.[14] The colonial state was, however, concerned not to appear entirely negligent. It recognized an obligation to 'avert death from starvation by the employment of all means practically open to the resources of the state', though this duty was to be discharged 'at the lowest cost compatible with the preservation of human life from wholesale destruction'.[15] It was often in order to demonstrate how few deaths were attributable to starvation, not how many—and to show that epidemic mortality

---

[12] Subsequently Carter slightly modified his views, acknowledging that the 1877 famine in Bombay was 'a sanitary catastrophe which will not soon be forgotten', but still arguing that the spread of relapsing fever was 'independent of local dearth'—H. Vandyke Carter, *Spirillum Fever: Synonyms: Famine or Relapsing Fever as Seen in Western India* (London: A. and J. Churchill, 1882), 15–25.

[13] F.W. Cragg, 'Relapsing Fever in the United Provinces of Agra and Oudh', *Indian Journal of Medical Research [IJMR]* 10, no. 1 (1922): 137.

[14] David Hall-Matthews, *Peasants, Famine and the State in Colonial Western India* (Basingstoke: Palgrave-Macmillan, 2005).

[15] 'Lord Lytton's Minute on Famine Policy, 12 August 1877', in *The Evolution of India and Pakistan, 1858–1947: Select Documents*, ed. C.H. Philips (London: Oxford University Press, 1962), 669.

was largely independent of hunger—that medical expertise was invoked.[16] Public health occupied a relatively low place in the hierarchy of state famine activity and expertise, with provincial famine codes placing medical aid and sanitary inspection far down the list of priorities.[17] Once famine was declared, the code required regular 'public health' reports from affected districts, but these tended to be remarkably brief and complacent in tone. A report on the 1899–1900 Punjab famine concluded, somewhat grudgingly, that 'the only way in which scarcity affected the health of the people was that so many were reduced by it that they were unable to withstand illnesses which under more fortunate circumstances might not have proved fatal'. The 'calamity caused by a poor season' was, the report continued, 'much ... to be regretted [but] cannot be prevented and can only be alleviated by means of ordinary medical and sanitary measures'.[18]

Even when it came, in the aftermath of famine, to explicit consideration of poverty, it was seldom medical expertise that prevailed. In 1888, a decade after large areas of India had endured one of the worst famines on record, the viceroy, Lord Dufferin, called for a report on 'the condition of the people of India', but in this exercise not a single medical officer was called upon to express an opinion. Since, like famine, the 'condition of the people' was seen as primarily a revenue issue, it was district collectors and other administrators who were consulted—and few of them recognized any causative link between poverty and disease.[19]

---

[16] For example, *Selection of Papers Relating to the Famine of 1896–97 in Bengal* (Calcutta: Bengal Secretariat Press, 1897), 2: 191.

[17] Jean Drèze, 'Famine Prevention in India', in *The Political Economy of Hunger*, ed. Jean Drèze and Amartya Sen (Oxford: Oxford University Press, 1990), 2: 26, n. 45.

[18] *The Punjab Famine of 1899–1900* (Lahore: Punjab Government Press, 1901), 1: 28.

[19] *Result of Enquiries Made in 1888 by Lord Dufferin into the Condition of the People of India* (London: HMSO, 1902). A similar optimism can be found in F.H.B. Skrine, *Memorandum on the Material Conditions of the Lower Orders in Bengal during the Ten Years from 1881–82 to 1891–92* (Calcutta: Bengal Secretariat Press, 1892), which concludes (p. 9) that, blessed with a warm climate and having few wants, 'the peasantry of Bengal are happy and prosperous'.

POVERTY AND THE NEW PUBLIC HEALTH

By the 1890s the colonial medical establishment in India was moving away from miasmatic theories of disease to explanations grounded in bacteriology, parasitology, and medical entomology. The new 'scientific' focus of this 'biomedical mission', as Hamlin describes it, tended to still further diminish rather than enhance the attention paid to poverty as a causative factor. [20] One example of this was the epidemic of bubonic plague that erupted in Bombay in 1896. [21] The common-sense view, as expressed in the press and initially by some government officers, was that plague was essentially 'a poor man's disease', for it seemed to be the poorer quarters of Bombay city that were hardest hit in the epidemic's opening phase. [22] But increasingly the poverty question was marginalized as the British responded to the epidemic by focussing, first, on the inspection of Indian bodies and then, as epidemiological knowledge progressed, on rats, rat fleas, and their extermination, and on anti-plague inoculation. The reports and recommendations of the Indian Plague Commission and the Plague Advisory Commission between 1900 and 1908 further directed expert opinion to the need for more detailed bacteriological and entomological studies. Eradicating rats became a far higher priority than eliminating poverty, though neither objectively appeared practicable.

The history of malaria followed a similar trajectory. Medical accounts of the disease by the early 1900s emphasized environmental factors, such as the waterlogged soils or urban localities in which anopheles mosquitoes bred: the solution accordingly lay in improved drainage,

[20]   Christopher Hamlin, 'Could You Starve to Death in England in 1839? The Chadwick-Farr Controversy and the Loss of the "Social" in Public Health', *American Journal of Public Health* 85, no. 6 (1995): 856.

[21]   For the context of this epidemic, see Myron Echenberg, *Plague Ports: The Global Urban Impact of Bubonic Plague, 1894–1901* (New York: New York University Press, 2007).

[22]   See J.A. Turner, Health Officer, to Municipal Commissioner, Bombay, 9 August 1905, India, Home (Sanitary), no. 329, February 1906, National Archives of India [NAI], New Delhi. According to S.H. Butler, Secretary, United Provinces, to Secretary, India, Home (Sanitary), 13 September 1905, J.A. Turner, no. 331. But not every physician agreed, even at the outset, that plague was 'a poor man's disease': see A.G. Viegas, *Bubonic Plague in Bombay* (Bombay: Tara-Vivechaka Press, 1897), 24.

the elimination of malaria-carrying mosquitoes, and the promotion of quinine prophylaxis.[23] Where there was a perceived link between malaria and poverty it was often in the inverted form, that by sapping human vitality, malaria caused poverty rather than vice versa. Patrick Hehir argued in 1927 that 'the eradication of malaria from India would, in a single generation, convert that country into one of the most prosperous in the world'.[24] Such a view, in theory, placed public health in a position of considerable authority, empowered to eradicate poverty by eliminating the diseases that caused it, but in actuality political and financial constraints, and the tendency for medical research to become increasingly laboratory-based and specialized, militated against such a hypothetical ascendency.

But there were medical and public health professionals for whom poverty did occupy a primary role. S.R. Christophers' pioneering investigation into the devastating malaria epidemic in Punjab in 1908, in which 300,000 people died in the space of two months, proposed on the basis of detailed statistical analysis a direct correlation between high food prices (as a proxy for economic distress) and heightened mortality among the poor. Christophers labelled this 'the human factor', though in later research he seemed to change his mind and reject a direct equation between poverty and malaria mortality. Why he did so remains a matter of debate.[25] Perhaps he came to realize that malaria was a more complex epidemiological phenomenon than he had at first imagined, subject to a host of different social and environmental, as well as economic, variables. Certainly, when the Imperial Malaria Conference convened at Simla in 1909 his economic argument was

[23] C.A. Bentley, *Malaria and Agriculture in Bengal: How to Reduce Malaria in Bengal by Irrigation* (Calcutta: Bengal Book Depot, 1925).

[24] Patrick Hehir, *Malaria in India* (London: Oxford University Press, 1927), 7–8.

[25] Sheila Zurbrigg, 'Re-thinking the "Human Factor" in Malaria Mortality: The Case of Punjab, 1868–1940', *Parassitologia* 36, nos 1–2 (1994): 121–35; Sheldon Watts, 'British Development Policies and Malaria in India, 1897–c.1929', *Past and Present* 165 (1999): 141–81; Christophers was not the first in India to use the expression 'the human factor': for example, B.B. Grayfoot, 'The Human Factor in the Spread of Plague, and the Lesson It Teaches', *Indian Medical Gazette* [*IMG*] 32 (May 1897): 163–5. But he gave it a new authority and a more economic significance.

strongly contested, and by the time he reported to the League of Nations' Malaria Commission in 1929, he had virtually abandoned the poverty hypothesis for a greater emphasis on environmental factors.[26]

Even though Ronald Ross and other leading malariologists in India recognized that the majority of fatalities occurred among 'the children of the poor', the general trend of malaria research in India showed the social understanding of poverty losing out to medical entomology and the dominance of 'biological expertise'.[27] The investigation begun by Christophers in Punjab in 1909 was transferred four years later to C.A. Gill, who expanded its scope beyond malaria to the consideration of other diseases, including plague and influenza, across India and beyond. In this enlarged enquiry Gill sought to develop a 'quantum theory' of epidemics in which poverty and the 'human factor' were ultimately assigned a relatively minor role.[28]

As indicated earlier, one of the few areas where medical and sanitary discourse did specifically address 'the government of poverty' was in the context of urban ill-health. Among the opening observations made in *McNally's Sanitary Handbook for India*, first published in Madras in 1889 and in its sixth revised edition by 1923, was the statement that poverty was 'one of the greatest bars to sanitary improvement'. Bad sanitary conditions were in themselves a cause of poverty, 'so that poverty and sickness react upon each other and tend to diminish the vital power of a people and their capacity for improvement'.[29] But, while acknowledging the underlying importance of poverty, works like *McNally's* did little to explicate the actual effects of poverty on health or to provide specific guidance as to how the vicious circle

---

[26] *Proceedings of the Imperial Malaria Conference Held at Simla in October 1909* (Simla: Government Central Branch Press, 1910); S.R. Christophers, 'Note on Malaria Research and Prevention in India', in *Report of the Malaria Commission on Its Study Tour in India* (Geneva: League of Nations, 1930).

[27] Ira Klein, 'Development and Death: Reinterpreting Malaria, Economics and Ecology in British India', *Modern Asian Studies* 38, no. 2 (2001): 155; D.J. Bradley, 'The Particular and the General: Issues of Specificity and Verticality in the History of Malaria Control', *Parassitologia* 40, nos 1–2 (1998): 6.

[28] Clifford Allchin Gill, *The Genesis of Epidemics and the Natural History of Disease* (London: Baillière, Tindal and Cox, 1928), 85–96.

[29] A.J.H. Russell, *McNally's Sanitary Handbook for India* (Madras: Superintendent, Government Press, 1923), 4.

between poverty and poor sanitation could be broken. One exception was in the burgeoning city of Bombay, where a series of three long-serving municipal health officers from the mid-1860s through to the 1920s identified poverty (along with related factors such as poor housing, inadequate diets and public ignorance of basic hygiene) as a major factor in persistently high levels of sickness and mortality. In a fairly typical statement, Bombay's second municipal health officer, T.S. Weir, wrote in his 1875 report on the factors that lay behind successive cholera epidemics in the city:

> The causes of these epidemics lie, I believe, in the poverty of the people, and the misery in which they live. The poverty of the great masses of the people of this city is incredible. Obliged to support large families, and compelled directly or indirectly to lay by a large proportion of their incomes for marriage and other ceremonies, they live on what I may without exaggeration call a starvation diet.[30]

Nor was it only cholera that prompted attention to poverty. In 1902, Weir's successor, J.A. Turner, reviewed at length factors responsible for persistently high levels of infant mortality in the city. His list included the physical weakness of working and under-age mothers, the inadequate nourishment of infants, and the insanitary and overcrowded surroundings in which so many of the city's poorer inhabitants lived, but the essence of his argument was poverty. 'I have stated above', he added, gesturing towards the relative powerlessness of his own profession, 'that poverty is at the root of the evil. This direct cause of much infant mortality is an influence with which the Municipality cannot deal, and it is out of the reach of remedies which we are competent to recommend.'[31]

[30] T.S. Weir, 'Annual Report of the Health Officer, Annual Report of the Municipal Commissioner of Bombay', 1875 (Bombay: Education Society's Press, 1876), 148. Weir repeated his view of the correlation between cholera and poverty almost twenty years later: Weir, 'Annual Report of the Health Officer', *Administration Report of the Municipal Commissioner for the City of Bombay, 1894–95* (Bombay: 'Times of India' Steam Press, 1895), 571.

[31] J.A. Turner, 'Report of the Executive Health Officer', *Administration Report of the Municipal Commissioner for the City of Bombay, 1902–03* (Bombay: 'Times of India' Press, 1903), 169. Further evidence of Turner's concern for the health of Bombay's poor can be seen in relation to the milk supply, urban overcrowding, plague, and tuberculosis: J.A. Turner, 'Report of the Executive

There may be several reasons why poverty commanded even this degree of prominence in Bombay's municipal health reports. Bombay was (as Mumbai still is) one of the largest concentrations of urban poverty and slum housing in South Asia. In the narrower arena of the city—and that in one of India's leading industrial and commercial cities and certainly one of its most affluent—the effects of poverty were more evident, and more readily quantifiable, than across entire provinces consisting mostly of small towns and villages.[32] Moreover, in Bombay sanitary and health measures were applied more resolutely and more systematically than almost anywhere in India (apart from the relatively privileged army and jails), and yet still failed to produce the intended results and stem the tide of sickness and death. As well as the exceptional vulnerability of a busy international port, an exposure which the advent of plague in 1896 and again of influenza in 1918 highlighted, poverty was one reason health officers could cogently advance to explain why greater success repeatedly eluded Bombay's urban sanitary order. At the same time, as Turner's remarks indicate, the city's health officers recognized that there was a limit to what they alone could do—without greater public awareness or more active support from industrialists and other leading employers. But the prominence given to poverty in the medical and sanitary reports on Bombay was not widely shared, even in other major cities like Calcutta and Madras.

Another instance of the growing, if erratic, attention given to poverty in medical and public health discourse was the influenza

---

Health Officer', *Administration Report of the Municipal Commissioner for the City of Bombay, 1904–05* (Bombay: Times Press, 1905), 187–9, 194–7; J.A. Turner, 'Report of the Executive Health Officer', *Administration Report of the Municipal Commissioner for the City of Bombay, 1915–16* (Bombay: Times Press, 1916), 16–19, 41–52.

[32] The visibility of poverty and its effects is a repeated trope in the Bombay health reports. In the first of these, in 1866, T.G. Hewlett remarked: 'It is only those persons who go into the houses of the poorer class in the native town who can picture the squalid poverty, the sallow unhealthy appearance of the inhabitants of some of the Tenant Houses.' *Bombay Municipality: Health Officer's Report, 1866*, IOR. It is ironic that Hewlett was later accused by Weir, his successor, of ignoring poverty as factor in the incidence of relapsing fever in the city in 1877.

pandemic of 1918–19. Although influenza caused heavy mortality around the world, India was exceptionally hard-hit by this disease, with an estimated 12–14 million deaths occurring in two main waves (of which the second was the more devastating, especially in rural areas) between June 1918 and the early months of 1919. Given a case mortality of around 10 per cent, at least 125 million people may have been stricken by the virus.[33] As with plague twenty-two years earlier, Bombay city was severely affected and in turn acted as an epicentre for the further dissemination of the disease.[34] While various factors were adduced to explain such staggering levels of mortality and morbidity (including the sheer inadequacy of existing public health services), it is striking how often poverty was cited as a crucial element, especially in triggering the pneumonia from which many influenza victims actually perished. While the existence of outright famine was not widely acknowledged, the end of the First World War was a period of high food prices and shortages across India, exacerbated by partial failure of the 1918 monsoon. As the interim report of the Government of India's Sanitary Commissioner put it in 1919, 'Actual famine conditions prevailed nowhere, though there was undoubtedly a greater degree of malnutrition than has been the case during many years past.'[35]

Subsequent comment from provincial health officers went still further in highlighting poverty. The Sanitary Commissioner for the United Provinces observed that, if influenza had proved far more fatal in India than in Western countries, 'it is due not to the fact that sanitary conditions are worse in India, but to the fact that economic conditions are worse. The people of India are worse housed, worse clothed and worse fed and can consequently offer less resistance to disease than those in England and America.' If such mortality were to be prevented in future, he added (perhaps in order to free his own service

---

[33] I.D. Mills, 'The 1918–1919 Influenza Pandemic: The Indian Experience', *Indian Economic and Social History Review* 23, no. 1 (1986): 1–40; R. Knowles, 'The Medical Aspects of the Indian Census of 1921', *IMG* 59 (September 1924): 466.

[34] Mridula Ramanna, 'Coping with the Influenza Pandemic: The Bombay Experience', in *The Spanish Influenza Pandemic of 1918–19: New Perspectives*, ed. Howard Phillips and David Killingray (London: Routledge, 2003), ch. 6.

[35] F. Norman White, *A Preliminary Report on the Influenza Pandemic of 1918 in India* (Simla: Government Monotype Press, 1919), 12.

from blame), 'the changes necessary to produce this result depend on changes in economic conditions and certainly not on the exertions of a sanitary staff'.[36] Bengal's Sanitary Commissioner adopted a similar line, in effect repeating what Christophers, his one-time collaborator, had written a decade earlier about the close relationship between high food prices and ill-health. 'The almost prohibitive price of cloth', wrote C.A. Bentley, '... greatly exaggerated the distress among the poorer classes. As a result of inadequate nourishment and clothing the vitality of the population generally was lowered and this contributed to the large death-rate from fever (including influenza).'[37] Although it would be a mistake to over-emphasize the lasting impact of the 1918–19 influenza epidemic on public health policy in India, it can be seen as one factor that provided poverty a heightened public and professional profile.

POVERTY, POPULATION, AND NUTRITION

During the interwar years, two aspects of the poverty–public health debate gained particular prominence and to some extent stood in opposition to each other. One was population, the other nutrition. Although economic, environmental, and even cultural explanations were widely invoked to explain the famines of the nineteenth century and the consequent heavy mortality, there had also been a strong undercurrent of Malthusianism to the way in which these recurrent episodes were discussed. 'Overpopulation' was blamed for India's poverty and hence its vulnerability to such 'natural checks' as famine and disease.[38] The question of acute poverty was thus construed as one of

[36] D. Mactaggart, 'Report on the Epidemic of Influenza in the United Provinces during 1918', *Annual Report of the Sanitary Commissioner of the United Provinces of Agra and Oudh, 1918* (Allahabad: Superintendent, Government Press, United Provinces, 1919), 15 A.

[37] Charles A. Bentley, *Report on Sanitation in Bengal, 1918* (Calcutta: Bengal Secretariat Book Depot, 1919), 9. See, too, W.H.C. Forster, 'The Influenza Epidemic of 1918 in the Punjab', *Report of the Sanitary Administration of the Punjab, 1918* (Lahore: Superintendent, Government Printing, Punjab, 1919), 16.

[38] S. Ambirajan, *Classical Political Economy and British Policy in India* (Cambridge: Cambridge University Press, 1978).

'excess' population. According to one official in the 1880s, constant pressure on the soil, accentuated by early marriages and the division of landholdings created, in the absence of sufficient industry and emigration to absorb it, an unsustainable number of 'almost paupers', until famine intervened and swept away 'its myriads'.[39] With the decline of famine, at least in its most extreme and recurrent form after 1908 (apart from the Bengal famine of 1943), Malthusianism passed out of favour, only to be resurrected in the late 1920s. The 'population problem' was given new urgency with the revelation in the 1931 census that India's population had increased by 10 per cent over the previous decade. This sparked fresh controversy as to whether India was 'overpopulated' and revived Malthusian claims that rapid demographic growth was the greatest issue India now faced.[40]

Malthusianism did not go uncontested. C.A. Gill declared in 1928 that Malthus was 'no longer numbered amongst the prophets'. His views, 'which recall the medieval attitude towards disease in general and epidemics in particular, have few advocates amongst men of science'. In Gill's view Malthus was a distraction from the real issues facing modern epidemiology.[41] But Gill's dismissal of Malthusianism proved premature. Already by the late 1920s 'the population problem' was being touted as the gravest economic and public health issue India faced, with dire parallels drawn with Ireland on the eve of the 1846 famine.[42] State doctors and public health officers (like Gill himself) played a prominent part in this debate and on both sides. Among the most strident figures involved was J.W.D. Megaw, who argued vehemently that medical intervention had inadvertently fuelled the crisis by reducing mortality from epidemic disease while little had been done to increase agricultural productivity or curb rampant population growth. 'Nature', he told an audience in London in 1934, 'now threatens to

[39] F.A. Nicholson, *Manual of the Coimbatore District* (Madras: Government Press, 1887), 257.

[40] David Arnold, 'Official Attitudes to Population, Birth Control and Reproductive Health in India, 1921–1946', in *Reproductive Health in India: History, Politics, Controversies*, ed. Sarah Hodges (Hyderabad: Orient Longman, 2006), 22–50.

[41] Gill, *Genesis of Epidemics*, ix–x.

[42] Editorial (probably by Megaw), 'The Population Problem in India', *IMG* 63 (July 1928): 328–9.

take her revenge for our interference with her destructive powers ...
we have upset her balance without applying a counterpoise.' Endemic
poverty was more consequence than cause and was largely of Indians'
own making since they had hitherto failed to adopt the birth-control
measures needed to check population growth and to prevent it still
further outstripping production.[43] Influential journals like the *Indian
Medical Gazette*, which Megaw for a time edited, threw their weight
behind the argument that population growth was now the foremost
problem India's public health regime had to face. Only birth control could prevent India's 350 million people from slipping into a still
greater catastrophe than that which engulfed Ireland's 8 million in the
1840s.[44]

The other key area of debate was in relation to diet, which could
be seen as an alternative route to Malthusianism in addressing India's
wider social and economic problems and one which saw the eradication
of poverty and malnutrition as of greater or more immediate importance than population growth and birth control.[45] 'Poverty', declared
the *Indian Medical Gazette* in 1922, 'is nowhere to be seen in more acute
form than in India.' It went on to say that millions of Indians lived in
what the American nutritionist E. V. McCollum had called 'the twilight
zone' where small shifts in diet could have critical effects on health.[46]
Scientific investigation of nutrition from the early 1920s onward began
to demonstrate clear links between poverty and ill-health. Initially,
however, nutritional deficiency (like poverty itself) was often seen in
India less as a consequence of economic deprivation than of custom
and culture, through religious taboos on certain foodstuffs, regional

[43] J.W.D. Megaw, 'Population and Health in India: The Real Problem',
Asiatic Review 30, no. 102 (1934): 250. For a similar view, see A.J.H. Russell
and K.C.K.E. Raja, 'The Population Problem in India', *IJMR* 23, no. 2 (1935):
545–68.

[44] Editorial (probably by R. Knowles), 'More about the Population
Problem', *IMG* 67 (March 1932): 153–7.

[45] David Arnold, 'The "Discovery" of Malnutrition and Diet in Colonial
India', *Indian Economic and Social History Review* 31, no. 1 (1994): 1–26; V.R.
Muraleedharan, 'Diet, Disease and Death in Colonial South India', *Economic
and Political Weekly* 29, nos 1–2 (1994): 55–63.

[46] 'The Economic Factor in Tropical Diseases', editorial, *IMG* 57
(September 1922): 341.

dietary preferences (as for rice over other more nutritious grains), the social subordination of women and hence the deficient nature of their diets, and so on—in other words, the perceived solution lay in changing cultural values rather than in tackling poverty.

An influential voice in this debate was Robert McCarrison of the Nutrition Research Laboratories at Coonoor. In the 1920s, McCarrison drew a stark dichotomy between 'well-constituted' 'Sikh' and 'Pathan' diets, consisting of wheat and dairy products, and the superior physique resulting from them, and the 'ill-constituted' diets of 'poor rice eaters' in Madras and Bengal and the weak races they produced. Using laboratory experiments with rats rather than field observation, McCarrison employed stereotypical images of 'martial' and 'un-martial' races, thereby endorsing earlier colonial thinking about the cultural determinants of health.[47] But McCarrison helped lay the foundations for nutrition research in India, and, as an effective publicist, garnered support (not least among Indians) for his belief in the centrality of nutrition to Indian well-being. Late in his career McCarrison shifted the emphasis in his thinking from cultural to economic factors. He still saw the diversity of races, and the consequent contrasts in diets and physique, as providing 'unrivalled material' for investigating nutrition in India. But he stressed more strongly the extent to which nutrition was linked to poverty as well. The nutrition researcher, he wrote, 'cannot ... lose sight of the economic aspect of his work which must be of service to the poor as well as to the rich'. The researcher should demonstrate not only what were the best diets but also those that were affordable even by the poor, 'for India is a land in which the poor far exceed the rich in numbers'.[48] McCarrison further argued that public health was a matter for the public to decide. The solution to India's nutrition problem did not lie with nutritionists alone. It was 'in the hands of others—the people of India themselves'.[49] A proper knowledge of nutrition could potentially empower the poor.

[47] R. McCarrison, 'A Good Diet and a Bad One: An Experimental Contrast', *IJMR* 14 (1927): 619–54. For an assessment of McCarrison's achievements, see James Vernon, *Hunger: A Modern History* (Cambridge, MA: Harvard University Press, 2007), 104–8.

[48] R. McCarrison, 'The Study of Nutrition in India', *Current Science* 1, no. 1 (1932): 3.

[49] McCarrison, 'The Study of Nutrition in India', 3.

The close of McCarrison's Indian career in the mid-1930s coincided with intense and unprecedented interest in the poverty question. This arose in part from a recognition that the decline of famine had not brought an end to the problem of endemic hunger, and also that, with the slow emergence of India's 'development regime',[50] mass poverty and the resulting ill-health and inefficiency of Indian workers stood in the way of material progress. The new critique of poverty came especially from left-wing nationalists, who saw an intimate connection between poverty, public health, and social justice. In 1929, Rajani Kanta Das argued that 'of all the important social, political and economic problems calling for immediate solution ... none is more complex and more difficult than that of the abject and perpetual poverty of the masses throughout the length and breadth of the country'. Das observed that, even with famine in abeyance, 'perpetual starvation' still stalked the land.[51] Another economist, N. Gangulee, similarly argued in 1939 that India had no more pressing problem than health and nutrition. For him the question of nutrition was closely bound up with the economic organization of society and the 'contradictions inherent in the capitalist mode of production'. He was scornful of claims that 'overpopulation' was the cause of Indian poverty but he was equally sceptical that cultural choice dictated poor diets. Only the eradication of poverty would solve the nutrition problem.[52]

In the West, widespread unemployment and the far-reaching effects of the Depression had generated a new concern for nutrition: in India, where unemployment was less of an issue, this translated into a new awareness of the impact of deficient diets on the health of the population as a whole and of the poorer classes and vulnerable groups like women and children. Crucial to this new poverty-focused view of public health was the work of W.R. Aykroyd, McCarrison's successor as Director of the Nutrition Research Laboratories. Aykroyd

---

[50] David Ludden, 'India's Development Regime', in *Colonialism and Culture*, ed. N.B. Dirks (Ann Arbor: Michigan University Press, 1992), 247–87.

[51] Rajani Kanta Das, 'The Problem of India's Poverty', *Modern Review* 46, no. 4 (1929): 365–70.

[52] N. Gangulee, *Health and Nutrition in India* (London: Faber and Faber, 1939).

was something of an exception in not being a member of the Indian Medical Service and so not burdened with the same prejudices and presumptions as those whose entire career had been spent in the Indian services. Highly regarded for his earlier work on nutrition for the League of Nations Health Organization, Aykroyd brought to nutritional science in India a much-needed empirical underpinning and a visionary understanding of how diet was reciprocally linked to poverty and integral to public health.[53] Aided by a team of committed Indian researchers, and financial support from the Indian Research Fund Association, Aykroyd oversaw a series of more than fifty food surveys by 1940.[54] These yielded the first field surveys of Indian diets, identified differences in consumption within households, and provided significant markers (such as body weight, calorific intake, and vitamin content) through which nutritional poverty could be accurately assessed. The surveys targeted social groups, especially poor urban and rural workers, women and children, who were most at risk from deficient diets, they directed attention to several hitherto little known but common deficiency diseases, and they made practical recommendations as to how even the poor could improve their daily diets.[55] Evidence from these studies began to inform government policy, as in Madras where free milk and cod-liver oil were introduced for schoolchildren

[53] Iris Borowy, *Coming to Terms with World Health: The League of Nations Health Organisation, 1921–1946* (Frankfurt-am-Main: Peter Lang, 2009), 382–7. Even the *Indian Medical Gazette*, often sceptical about the new nutritional science, hailed Aykroyd's work for the League of Nations and recognized that poverty was 'undoubtedly a serious complicating factor' in the incidence of disease: 'Nutrition and Public Health', editorial, *IMG* 70 (November 1935): 632. A year later the journal modified still further its earlier stance on India's 'Malthusian nightmares', 'The Population Problem', editorial, *IMG* 71 (May 1936): 278.

[54] W.R. Aykroyd, *Note on the Results of Diet Surveys in India* (New Delhi: Indian Research Fund Association, 1939).

[55] W.R. Aykroyd and B.G. Krishnan, 'The Carotene and Vitamin A Requirements of Children', *IJMR* 23, no. 3 (1936): 741–5; Aykroyd and Krishnan, 'The Deficiencies of the South Indian Diet', *IJMR* 25, no. 2 (1937): 367–72; W.R. Aykroyd, K.B. Madhava, and K. Rajagopal, 'The Detection of Malnutrition by Measurements of Arm, Chest, and Hip', *IJMR* 26, no. 1 (1938): 55–94.

in 1936.[56] At the all-India level, too, nutrition was becoming a major policy issue.[57]

Although the Second World War brought the diet surveys to a halt, it provided fresh impetus for a new nutrition policy. The government was forced to jettison laissez-faire and intervene on an unprecedented scale to regulate the grain trade and food consumption. Publication of the Beveridge Report in Britain in 1942 gave further impetus to the social objectives of public health in India. This was reflected in the Health Survey and Development Committee's report in 1946, which in its discussion of the 'social background of ill-health' quoted Beveridge enthusiastically even while adding apologetically that it was beyond the scope of the committee 'to suggest ways and means by which poverty and unemployment should be eliminated'.[58]

Famine had not ceased to be a factor in the understanding of mass poverty. The Bengal famine of 1943 and the official inquiry it spawned strengthened the belief that the quality as well as the quantity of foodstuffs consumed by the public was a cause for public concern and an inducement to state action. The report of the Famine Inquiry Commission (of which Aykroyd was a member) in 1945 contained an extended discussion of 'population, nutrition and food policy'. Having attacked the 'fanatical Malthusian' position, the report argued instead that nutrition was vital to public health and that income levels needed to rise in order to allow an improvement in the diets of the masses. The commission declared that 'a well-balance and satisfactory diet' was currently 'beyond the means of a large section of the population' and that this was a prime cause of debility, disease, and death. It followed that the 'improvement of nutrition' was an 'essential part of [any] public health programme in India'.[59] Seldom had the 'government of poverty' enjoyed such prominent endorsement.

But the science of nutrition and the agency of public health were by no means the only ways in which the problem of poverty could

[56] Local Self-Government (Public Health), no. 2649, 23 December 1936, Tamil Nadu Archives, Chennai.

[57] Education, Health and Lands (Health), 37-6/37, 1939, NAI.

[58] *Report of the Health Survey and Development Committee* (Delhi: Manager of Publications, 1946), 1–17.

[59] *The Famine Inquiry Commission: Final Report* (Madras: Government Press, 1945).

be addressed. One alternative came from the physicist Meghnad Saha, who in the 1930s proposed that the dual problem of flooding and malaria in Bengal could be resolved by large-scale measures of river-control and irrigation. Saha made clear that poverty eradication lay at the core of his proposals, arguing in 1938, 'If we desire to fight successfully the scourge of poverty and want from which 90% of our countrymen are suffering ... we must make the fullest use of the power which a knowledge of Nature has given us.'[60] The agency of public health was, it seems, not adequate to the task. Saha's scheme, inspired by Soviet and New Deal precedents, implied the massive deployment of state resources, and underlay the formation of the National Planning Committee in 1938.[61] Saha's ambitious schema helped redirect attention towards wholesale intervention by the state and away from the small-scale, self-help governmentality that diet surveys and nutrition science had tried to encourage. His vision gave priority to science and technology—the engineering needed to build big dams, generate electricity, and regulate river flows—and defined a developing tendency to look to science and technology rather than to medicine and public health to resolve India's poverty problem. Public health and state engineering were not necessarily rivals: they could indeed be complementary approaches. But the fact that engineering of the kind and scale envisaged by Saha was being invoked as a primary means of removing poverty suggested the inability of public health to perform that monumental task.

[60] M.N. Saha, 'The Problem of Indian Rivers (1938)', in *Collected Works of Meghnad Saha*, ed. Santimay Chatterjee (Bombay: Orient Longman, 1986), 2: 64–91.

[61] On Saha's vision and influence, see Shiv Visvanathan, *Organizing for Science: The Making of an Industrial Research Laboratory* (Delhi: Oxford University Press, 1985), ch. 3.

2 Tubercular Optics

*Health, Techno-science, and the Obfuscation of Poverty**

Lakshmi Kutty

That tuberculosis (TB) is a social disease, quintessentially a disease related to poverty, hardly needs re-stating. Literature on the decline and resurgence of TB in different regions and periods of time has noted its close relationship with standards of living as impacted by social and structural change, including wage rates and incomes, diet and nutrition, living and working conditions, urbanization, and health infrastructure.[1] Different meanings of poverty have the potential to

* I am grateful to Sarah Hodges and Mohan Rao for their support and encouragement, and especially to Mohan Rao for comments on earlier drafts. I thank Ramila Bisht, David Arnold, and participants at a workshop of this project for their responses on an earlier draft. I am also grateful to Emma Rothschild and Abhijit Banerjee for the opportunity to discuss this paper at the 'Economic History of Poverty' conference under the aegis of the Joint Centre for History and Economics, Harvard University, and the University of Cambridge.

[1] Thomas McKeown, R.G. Brown, and R.G. Record, 'An Interpretation of the Modern Rise of Population in Europe', *Population Studies* 26, no. 3 (1971): 345–82; D.P.S. Spence, J. Hotchkiss, C.S.D. Williams, and P.D.O. Davies, 'Tuberculosis and Poverty', *British Medical Journal* 307, no. 6907

foreground vastly different understandings of TB illness, treatment, and policy making, and with it enable differing possibilities of health and well-being for patients. Poverty has been understood as the lack of economic and social resources necessary for assuring health free from, or recovery from, TB illness.[2] It has been understood as inequality and exclusion from socioeconomic stability and political power, which leaves the poor more vulnerable to the disease and the possibilities of their treatment and healing fraught with uncertainty.[3] It has been deployed as a code for patient non-compliance or 'insufficient' patient-hood, warranting additional surveillance.[4] It has been understood as a function of the political economy of a state or region driven by global inequality and health sector reforms.[5]

India's National Tuberculosis Programme (NTP) was formulated in the 1960s in the backdrop of epidemiological surveys that identified TB as a public health problem nestled in resource constraints favourable to the spread of disease, such as poverty, malnutrition, overcrowding, increasing industrialization, seasonal rural unemployment,

(1993): 759–61; Rene and Jean Dubos, *The White Plague: Tuberculosis, Man and Society* (Boston: Little, Brown and Company, 1952); D. Banerji and S. Andersen, 'Sociological Study of Awareness of Symptoms among Persons with Pulmonary Tuberculosis', *Bulletin of the World Health Organization* 29, no. 5 (1963): 665–83; Amy L. Fairchild and Gerald M. Oppenheimer, 'Public Health Nihilism vs Pragmatism: History, Politics, and the Control of Tuberculosis', *American Journal of Public Health* 88, no. 7 (1998): 1105–17.

[2] M. Muniyandi, R. Ramachandran, P.G. Gopi, V. Chandrasekaran, R. Subramani, K. Sadacharam, et al., 'The Prevalence of Tuberculosis in Different Economic Strata: A Community Survey from South India', *The International Journal of Tuberculosis and Lung Disease* 11, no. 9 (2007): 1042–45.

[3] P. Farmer, *Infections and Inequalities: The Modern Plagues* (Berkeley: University of California Press, 2001).

[4] J. Porter and J. Ogden, 'Public Health, Ethics, and Tuberculosis: Is DOTs a Breakthrough or Inappropriate Strategy in the Indian Context?', *Indian Journal of Medical Ethics* 7, no. 3 (1999): 79–84.

[5] E. Jaramillo, 'Encompassing Treatment with Prevention: The Path for a Lasting Control of Tuberculosis', *Social Science & Medicine* 49, no. 3 (1999): 393–404; Imrana Qadeer, Kasturi Sen, and K.R. Nayar, eds, *Public Health and the Poverty of Reforms: The South Asian Predicament* (New Delhi: Sage Publications, 2001).

and migration.[6] These were overwhelmingly large and multifaceted problems calling for long-term structural change; they also called for scientific research aimed to address the immediate burden of disease and suffering in the population. This situation of lack of resources formed the impetus for research on novel scientific technologies and drug regimens that would rise above the socioeconomic contexts of impoverished patient populations. Research advances made with financial and scientific assistance from international organizations created a confidence that focusing on chemotherapeutics, technical tools, and programme organization and planning would address the problem of TB in the population.[7] Thus, even though TB policy formulation emerged, in India as elsewhere, from an awareness of the TB-poverty linkages, and these links formed the ostensible rationale behind the further development of TB policy, the question of poverty remained inadequately addressed.

What resulted from this confidence in scientific advance was a re-casting and re-coding of the association between TB and poverty. The most potent of these was a move away from questions of a political nature to questions of a techno-scientific/administrative/managerial nature. Apart from understanding poverty from the perspective of socioeconomic and political determinants that impact TB, as discussed earlier, poverty may be read in another way. It may be read as the thing being re-cast, re-coded through a techno-scientific framing, which nearly eliminates the social–structural–political meanings and potential attached to it and imbues it with a technical visage. By this re-casting it becomes possible to still address the problem of poverty, as earlier, but this time in a renewed way, and with the power and capacity to respond with technical solutions that will fit the problem. This chapter will explore this techno-scientific framing, which has re-cast the problem of poverty in TB into something that now merits technical tools for problem solving. It will elaborate the way this framing has been constructed and has attained legitimacy, and the way it has

---

[6] Indian Council of Medical Research, *Tuberculosis in India: A Sample Survey, 1955–58* (New Delhi: Indian Council of Medical Research, 1959).

[7] Tuberculosis Chemotherapy Centre, Madras, 'A Concurrent Comparison of Home and Sanatorium Treatment of Pulmonary Tuberculosis in South India', *Bulletin of the World Health Organization* 21, no. 1 (1959): 51–144.

been articulated through an individualized, behavioural framing of the structure of cure, as well as a cost-effectiveness framing of the attainment of health.

Drawing from a layered conceptual framework, the chapter will discuss two tools from the field of TB programme and policy: the development of the concept of *default* (irregularity in treatment completion, 'non-compliant' behaviour, 'inadequate motivation' in adhering to regimens) and techniques to arrest it; and of *Annual Risk of Infection* and the related *Styblo rule* (a statistical estimate of the probability of an individual acquiring new tuberculous infection or reinfection over a period of one year, and a related estimate of new smear-positive TB disease in a community). Smear-positive TB is pulmonary TB that tests positive for the presence of tubercle bacilli on a sputum smear test. Smear-positive pulmonary TB is infectious, and efforts for control of TB focus intensely on this category of patients.

RE-CASTING POVERTY THROUGH
TECHNO-SCIENTIFIC RATIONALITY

The domain of policy making is an intertwining of technical, managerial, and political characteristics; it is enmeshed in the power relations and hierarchies between citizens, experts, and political authorities. Such a reading has been interpreted as advancing an instance of Foucault's concept of 'political technologies'.[8] Dreyfus and Rabinow, dialoguing with Foucault, note that '[p]olitical technologies advance by taking what is essentially a political problem, removing it from the realm of political discourse, and recasting it into the neutral language of science. Once this is accomplished, the problems have become technical ones for specialists to debate'.[9] Alongside, there is also the question of the nature of scientific expertise, and how 'the power of the professions depends, at least in part, on the ability to make claims successfully about the scientific value of their work and the way in which their

[8] J. Keeley and I. Scoones, 'Understanding Environmental Policy Processes: A Review', IDS Working Paper 89 (1999), 5.

[9] H. Dreyfus and P. Rabinow, *Michel Foucault: Beyond Structuralism and Hermeneutics* (Brighton: Harvester, 1982), 196.

professional knowledge is grounded in precise, accurate and reliable scientific information'.[10]

Critics have, however, pointed out that given the unstable and dynamic nature of scientific judgements, such judgements tend to be 'artefacts of the chosen methodologies rather than [...] representations of reality'.[11] The validation of scientific judgements from being 'backed up by an organizational infrastructure' merits further scrutiny, as organizational backing tends also to 'emphasize those aspects of risk that its scientific bureaucracy has the tools to measure (expected losses) at the expense of less easily measured, but not necessarily less important, aspects of risk-bearing'.[12] Further, policies removed from the realm of politics can appear as 'technical responses to particular problems, which fall within the mandate or objectives for which an international organization has been established'.[13]

Thus, we have critical analyses of the controversies attending the development of international BCG vaccination policies by the World Health Organization (WHO) during the 1940s–1980s.[14] Additionally, there are studies of the creation between the 1960s and 1990s of the global technical strategy of Directly Observed Treatment, Short course (DOTS).[15] We have studies of the contending perspectives that marked research and policy directions for TB in two institutes of South India

---

[10] Bryan Turner, *Medical Power and Social Knowledge* (London: Sage Publications, 1995), 208.

[11] Donald T. Hornstein, 'Reclaiming Environmental Law: A Normative Critique of Comparative Risk Analysis', *Columbia Law Review* 92, no. 3 (1992): 573.

[12] Hornstein, 'Reclaiming Environmental Law', 573–5.

[13] Gill Walt, Louisiana Lush, and Jessica Ogden, 'International Organizations in Transfer of Infectious Diseases: Iterative Loops of Adoption, Adaptation, and Marketing', *Governance: An International Journal of Policy, Administration, and Institutions* 17, no. 2 (2004): 189–210, 190.

[14] Niels Brimnes, 'BCG Vaccination and the WHO's Global Strategy for Tuberculosis Control, 1948–1983', *Social Science and Medicine* 67, no. 5 (2008): 863–73.

[15] Jessica Ogden, Gill Walt, and Louisiana Lush, 'The Politics of "Branding" in Policy Transfer: The Case of DOTS for Tuberculosis Control', *Social Science and Medicine* 57, no. 1 (2003): 179–88; Walt, Lush, and Ogden, 'International Organizations in Transfer of Infectious Diseases'.

in the 1950s and 1960s.[16] These have all demonstrated the processes by which optimum scientific tools or policy proposals get identified as such and then promoted to various constituencies as international scientific 'best practice'. These have revealed not only the strategic roles that influential actors, 'epistemic communities', or organizations play in constructing and promoting specific facets of scientific knowledge, but also that the adoption of these 'best practice' models habitually displays the intermingling of several processes beyond plain scientific.[17]

Discussing the broad contours of TB health policy research in the early 2000s, Porter and Kielmann note that the contemporary international perspective in TB is the 'biomedical scientific model of infectious disease control'. It is 'based on a structure and system that has as its core a rational approach to policy making. Programmes are based on a biomedical approach with the use of rational processes to design, implement, and evaluate information. This is the "normal" perspective of the "technical advisors".'[18] The range of existing rational technical tools for identifying optimum policy proposals for reforming the health system include 'burden of disease assessments; evaluation of the evidence on the effectiveness and cost-effectiveness of interventions; analyses of health systems performance and available finances; and a community survey and political mapping exercise to assess the political feasibility of the different options'.[19] The trajectory of development of technical tools for TB control suggests that this contemporary model does echo comparable perspectives from at least the 1960s, if not earlier.

Rational choice approaches draw upon economic considerations of decision making and rational action.[20] Indeed the logics of political decision making in health care have often been intertwined with

[16] Sunil Amrith, 'In Search of a "Magic Bullet" for Tuberculosis: South India and Beyond, 1955–1965', *Social History of Medicine* 17, no. 1 (2004): 113–30.

[17] Ogden, Walt, and Lush, 'The Politics of "Branding"'.

[18] John Porter and Karina Kielmann, 'TB Control in India: The Need for Research in Policy and Decision-Making', *Health Administrator* 15, nos 1–2 (2003): 143–8, 144.

[19] Frenk 1995, Brugha and Varvasovszky 2000, cited in Porter and Kielmann, 'TB Control in India', 146.

[20] Jens O. Zinn, 'Recent Developments in Sociology of Risk and Uncertainty', *Forum: Qualitative Social Research* 7, no. 1 (2006), art. 30, http://nbn-resolving.de/urn:nbn:de:0114-fqs0601301.

the tools of health economics.[21] That is to say, assessments of rational choice are expected to weave into the demands of cost-benefit analyses that the market dictates. Attempts to combine rationality with pragmatic tool development call not only for calculation of the relationship between costs and benefits, but also suggestions that new initiatives should be adopted only if the expected benefits exceed the costs; the rational scientific choice is therefore one which maximizes the value of outcomes while minimizing any possibility of risk. Among the range of technical tools at the core of a rational approach to scientific research in health care, cost-effectiveness analysis is a defining criterion for programme planning, especially in contexts of poverty. It has, however, been critically examined for its inconsistency with considerations for improved health care and well-being for the poor attended by wider socioeconomic and political justice.[22]

In a particularly strong critique of the logic of cost-effectiveness, which had been used in the mid-1990s to argue that treatment of multidrug resistant TB (MDR-TB) was not feasible in resource-poor settings, Yong Kim et al. note that such a logic ignores and conceals the multiple social, political, economic, epidemiological, and pathophysiological factors that determine or constrain the access of the poor to health services. Foregrounding the impact of neo-liberal economic reforms and the burden of debt repayments on poor country health budgets during the period of structural adjustment that spanned the 1970s through the 1990s, they point out:

> Rather than first arguing for debt relief and the restoration of health care budgets as a condition for canceling debt, many health policy makers, including those in the TB world, simply looked to cost-effectiveness analyses to help them find and recommend the most favorable interventions within the context of 'limited' and in many cases shrinking resources.[23]

[21] Tiago Moreira, 'How to Investigate the Temporalities of Health', *Forum: Qualitative Social Research* 8, no. 1 (2007), art. 13, http://nbn-resolving.de/urn:nbn:de:0114-fqs0701139.

[22] Jaramillo, 'Encompassing Treatment with Prevention', 393–404.

[23] Jim Yong Kim, Aaron Shakow, Kedar Mate, Chris Vanderwarker, Rajesh Gupta, and Paul Farmer, 'Limited Good and Limited Vision: Multidrug-Resistant Tuberculosis and Global Health Policy', *Social Science & Medicine* 61, no. 4 (2005): 847–59, 854, doi:10.1016/j.socscimed.2004.08.046, http://www.ncbi.nlm.nih.gov/pubmed/15896895.

In a perspective rarely articulated in the field, they show clearly the ways in which cost-effectiveness logic was driven by falsely high drug pricing regimes, and how the transnational political economy of competing mandates among international aid organizations simply denied access to MDR-TB care to those in need.

Another dominant claim of a rational approach to scientific research relies on the way this is harnessed to the idea of risk. Risk is animated not only in its empiric form, as the possibility of an adverse outcome, but more in terms of being a paradigmatic construct wherein measurements and comparative assessments of risk, management of and safety mechanisms against risk, and so on, drive decision making in a majority of spheres.[24] Moreover, drawing from stages in the pathogenesis of TB, 'risk factors' have played a fundamental role in the development of conceptual and methodological tools central to our understanding of the epidemiology of TB.[25] The growing influence of risk factor epidemiology during the 1950s–60s displaced the specific cause model of disease and factored in the multiplicity of risk factors. However, this analysis remained confined to the individual level, and thus strategies for prevention and treatment highlighted change in individual behavioural aspects.[26] Much of the focus of public health action converged on 'seeking solutions to major health risks by urging individual responsibility and personal health action as compared with social and environmental remedies that address key health risks at their source'.[27]

Evolving from such a context, expert knowledge in TB sought to perfect the programmatic aspect of risk management such that among the complex web of conditions of inequality which mark TB, 'patient risk behaviour', default, and its surveillance came to play the most vital role in combating the disease. This often culminated in a tendency

[24] Hazel Kemshall, *Risk, Social Policy and Welfare* (Buckingham: Open University Press, 2002).

[25] Dipanjan Roy and L.S. Chauhan, 'Epidemiology of Tuberculosis', in *Tuberculosis Control in India*, ed. S.P. Agarwal and L.S. Chauhan (New Delhi: Directorate General of Health Services, Ministry of Health and Family Welfare, 2005).

[26] Mervyn Susser and Ezra Susser, 'Choosing a Future for Epidemiology: I. Eras and Paradigms', *American Journal of Public Health* 86, no. 5 (1996): 668–73.

[27] D. Mechanic, 'Changing Fortunes of Medical Sociology', *Social Science and Medicine* 36, no. 2 (1993): 95–102, 97.

to lay the responsibility for programme failure at the doorstep of the individual patient, seeking then to provide a combination of supportive and coercive actions to arrest default. Risk management is also operationalized through demands for universal standardization of concepts, techniques, and methodologies, each having the potential to be transported/replicated and to generate comparable data across regions. In such scenarios, however, there always exists the problem of interpretation of the evidence on risk, echoing critiques noting the unstable and dynamic nature of scientific knowledge.[28]

Porter and Kielmann note that processes surrounding the development of TB policy have been part of specific frameworks, structures, and systems, and policy decision-making has been an inherently political process. They argue that '[i]n order to link the technical with the political, it is therefore appropriate that we interrogate our approaches to the creation of TB programmes and their implementation with perspectives that go beyond the rational model of decision-making'.[29] The discussion in the following two sections will illustrate the manner in which a certain scientific rationality was articulated through the concepts of default and the tools of annual risk of tuberculosis infection (ARI) and the Styblo rule. It will elaborate the deployment of a cost-effectiveness rationale to effect the global legitimization of these two tools and generate a renewed scientific knowledge base for TB control, which, it was claimed, would better address the gravity of the TB situation in developing countries in which the combination of levels of poverty and burden of disease and death from TB was staggering.

'DEFAULT' AND SUPERVISED THERAPY: CENTERING PATIENT BEHAVIOUR IN THE TREATMENT STRUCTURE

From the time default/noncompliance was identified as an important factor in treatment, it singlehandedly steered the direction of global epidemiological research and programme planning in TB for nearly the

---

[28] Paul Higgs, 'Risk, Governmentality and the Reconceptualization of Citizenship', in *Modernity, Medicine and Health: Medical Sociology Towards 2000*, ed. Graham Scambler and Paul Higgs (London: Routledge, 1998).

[29] Porter and Kielmann, 'TB Control in India', 144–5.

next three decades; this was a seminal conceptual tool of the TB programme. As one of the sites developing research and programme tools that would give direction to global TB policy, Indian scientific institutes of the 1950s, the Tuberculosis Chemotherapy Centre (TCC), Madras, and the National Tuberculosis Institute (NTI), Bangalore, were set up as collaborative enterprises with the WHO and British Medical Research Council (BMRC). The BMRC's Tuberculosis Research Unit conceived and set up TCC in 1956 to study the applicability of ambulatory/ domiciliary (as opposed to sanatorium/hospital based) therapy in mass programme settings.[30] Additionally, the WHO helped set up the NTI to concentrate essentially on training district teams, the operational problems of programme application, and on epidemiology.[31]

A third important scientific body was the Paris-based International Union Against Tuberculosis (IUAT), a specialized non-governmental voluntary organization for global advocacy and scientific research in TB. The IUAT focused on developing models for national TB programmes across the world. An important part of its mandate was to set up national TB associations to assist governments of different countries in their anti-TB efforts, and exert pressure to influence political decision-making, as it saw government bodies bound within certain structures which prevented them from taking all the steps needed to combat the threat of TB.[32]

For the next few decades, several controlled trials, chemotherapy regimen studies, and epidemiological surveys were conducted among patients accessing, or identified by, TCC and NTI. Their research mandates were defined by the pressing demands and issues arising from the Indian context—lack of adequate resources, low living standards, fewer sanatoria, and hospital beds compared to the large numbers requiring TB treatment, expensive and prolonged therapy regimens, a majority

[30] Wallace Fox, Gordon Ellard, and Denis A. Mitchison, 'Studies on the Treatment of Tuberculosis Undertaken by the British Medical Research Council Tuberculosis Units, 1946–1986, with Relevant Subsequent Publications', *International Journal of Tuberculosis and Lung Disease* 3, no. 10 (1999): S231–S279.

[31] Wallace Fox, 'Tuberculosis in India: Past, Present and Future', *Indian Journal of Tuberculosis* 37, no. 4 (1990): 175–213.

[32] 'The Mutual Assistance Programme of the International Union against Tuberculosis', *Bulletin of the International Union against Tuberculosis* 33, no. 2 (1963): 116–35.

of which were out of reach of most patients—but the urgency was to gain insights for TB control across a range of developing countries. The disparity between the prevalence of TB and the availability of hospital beds, and of course the costs involved, made ambulatory care the only viable option for developing countries like India during this period. Of the various problems that plagued treatment in a context of inadequate resources, BMRC and IUAT studies in India and East Africa respectively played definitive roles in identifying and centering the concepts of irregularity in drug intake and default for further research and programme management.

BMRC–TCC Research Directions

Studies carried out in TCC, under the directorship of Wallace Fox, on home/ambulatory versus sanatorium care noted early on that compliance to drug regimens was a problem among ambulatory patients.[33] Fox started research on the potential efficacy of supervised therapy in TCC by the late 1950s.[34] Detailing the kinds of efforts taken to ensure compliance to therapy at TCC, Fox writes of a highly coordinated approach to monitor patients, which included

> (1) the involvement of the family as well as the patient during the pre-treatment diagnosis and assessment period, (2) frequently repeating during therapy the need to take the medicaments regularly and emphasizing the family's or a neighbour's role in supervision, (3) a policy of restricting the medication to antituberculosis chemotherapy whenever possible, (4) surprise visits to the home to (a) check the patient's stock of pills and (b) to collect a urine specimen to test for antituberculosis drugs, and (5) taking speedy absconder action if a patient failed to attend when due.[35]

He elaborates the exact modalities of this last point further in a footnote:

[33] Wallace Fox, 1958, cited in Ronald Bayer and David Wilkinson, 'Directly Observed Therapy for Tuberculosis: History of an Idea', *Lancet* 345, no. 8964 (1995): 1545–8.

[34] Tuberculosis Chemotherapy Centre (TCC), 'A Concurrent Comparison of Intermittent (twice weekly) Isoniazid plus Streptomycin and Daily Isoniazid plus PAS in the Domiciliary Treatment of Pulmonary Tuberculosis', *Bulletin of the World Health Organization* 31 (1964): 247–71.

[35] Fox, 'Tuberculosis in India', 188.

For every patient admitted to treatment a full list of addresses was obtained, namely 1) the patient's home address, 2) the addresses of his relatives and friends in Madras city, and 3) how often he visited them, 4) if employed, the place of work, 5) if children were at school, its address, 6) the address of his native place. If a patient failed to attend and the home was locked and the neighbours did not know where the patient or the family was, a systematic approach to the above alternatives was made until the patient was traced. (Adequate transport and devoted trained home visitors for this purpose were available.)[36]

Patients had to face many difficulties to undergo such treatment.[37] Nevertheless, by the early 1960s, Fox advocated that long term supervised therapy could be organized in special circumstances, even in developing countries.[38] This intense emphasis on a single aspect of the treatment modality, defaulting behaviour, led to a surge of research studies in two related areas—supervised therapy, and intermittent therapy. The second area of research emerged directly from the first, though this was not driven solely by pharmacological insights; studies on supervised chemotherapy started exploring the development of intermittent regimens 'since these would ease the task of supervision'.[39]

The era of short course chemotherapy emerged later, by the 1970s, in response to standard regimens (12 or more months, both daily and intermittent) failing because they were 'beyond the organizational and financial resources of the health services of nearly all these countries'.[40] The shortage was related not just to drugs, but also to the fact that

[36] Fox, 'Tuberculosis in India', 188n.

[37] Bayer and Wilkinson, 'Directly Observed Therapy for Tuberculosis', 1545–8; Wallace Fox, 'Self-Administration of Medicaments: A Review of Published Work and a Study of the Problems', *Bulletin of the International Union Against Tuberculosis* 32 (1962): 307–31; Amrith, 'In Search of a "Magic Bullet" for Tuberculosis', 113–30.

[38] Fox, 'Self-Administration of Medicaments', 307–31; TCC, 'A Concurrent Comparison of Intermittent (twice weekly) Isoniazid plus Streptomycin and Daily Isoniazid plus PAS in the Domiciliary Treatment of Pulmonary Tuberculosis'.

[39] Fox, Ellard, and Mitchison, 'Studies on the Treatment of Tuberculosis', S237.

[40] Wallace Fox, 'Compliance of Patients and Physicians: Experience and Lessons from Tuberculosis-Part II', *British Medical Journal* 287, no. 6385 (1983): 101–5.

these countries were not in a position to set up the infrastructural and personnel requirements for supervised treatment. When short course regimens were combined with the principle of intermittence this represented the highest advance in fully supervised chemotherapy.[41]

By the end of the 1960s–early 1970s BMRC research in various locations had concluded that 'effective treatment required direct supervision of therapy and that only such an approach could interrupt a general tendency [...] on the part of patients to cease taking medications when they no longer felt ill'.[42] This representation of patients as having a 'general tendency' to be non-compliant existed in spite of BMRC-TCC research being fully aware of the extent of poor patients' circumstances of poverty and deprivation. Many patients in the landmark home versus sanatorium study had a poor diet, less rest, inferior and crowded accommodation, insecure employment, and less nursing care.[43] Additionally, Fox noted that in a city with poor transport services where 'distances of up to five miles to and from the clinic have to be travelled[,] [s]ome of them receive limited financial assistance for their fares, but it is certain that many use their fare money to supplement their diet and are prepared to walk daily for their treatment'.[44]

In situations of limited resources, the research impetus, then, was to find treatment techniques that could bypass patients' circumstances of poverty through contracted solutions, not necessarily engage with the complexity of their socioeconomic contexts. What this did was to cement forever the focus of scientific responses to TB on behavioural and motivational aspects of the TB patient's encounter with the treatment. Newer developments in scientific technology would not diminish the individualized surveillance mode of the treatment regimen that had been charted out, and the public health system for TB as it shaped over the next few decades seemed to have few other ways to respond to the fragile status of TB patients in their larger contexts of deprivation.

[41] Fox, Ellard, and Mitchison, 'Studies on the Treatment of Tuberculosis'.

[42] Bayer and Wilkinson, 'Directly Observed Therapy for Tuberculosis', 1546.

[43] Tuberculosis Chemotherapy Centre, 'Concurrent Comparison of Home', 51–144.

[44] Fox, 'Self-Administration of Medicaments', 321.

Research Directions at NTI

As elsewhere, non adherence to drug regimens continued to plague Indian health administrators and research communities around the time of the NTP in 1962. Observations on defaulting behaviour often held patient ignorance about the importance of completing the treatment duration, discontinuation of drugs on feeling better, or laziness and indolence responsible for treatment failures. A 1963 study at NTI by Stig Andersen and Debabar Banerji was a notable contrast from studies done till then on TB; it positioned the patient as first and foremost a social entity, and acknowledged that patients' experiences are constituted by the social-structural space they occupy.[45] It focused attention on the experience of worry and suffering caused by ill health; it questioned the traditional definition of a treatment defaulter as too technocentric, arbitrary, and value-loaded; and provided relevant sociological insights into why patients were unable to maintain prolonged drug regimens. It pointed out that patient indifference to the dictates of treatment could be because 'the suffering due to tuberculosis was not strong enough to motivate him to continue treatment; this could be due to the presence of some other more pressing problems (for example, problem of hunger) overshadowing the problem of pulmonary tuberculosis'.[46]

While this study was a hopeful sign of the ways in which patient-oriented perspectives could drive research, it did not, however, radically shift attention away from patient default as a central issue of programme management. On the contrary, while proposing a new understanding of sociological factors underlying TB therapy, and the need to solve the problem of default, the study criticized treatment organization failures, administrative and managerial lacunae, and improper implementation of service delivery.[47] Instead, success for the NTP was to emerge from an overhaul of the 'slippery slope of sloppy treatment

[45] Banerji and Andersen, 'Sociological Study of Awareness', 665–83.

[46] D. Banerji, 'Tuberculosis as a Problem of Social Planning in India', *National Institute of Health Administration and Education Bulletin* 4, no. 2 (1971): 9–25.

[47] D. Banerji, 'Behaviour of Tuberculous Patients towards a Treatment Organisation Offering Limited Supervision', *Indian Journal of Tuberculosis* 14, no. 3 (1967): 156–72.

organization'.[48] The NTI Newsletter, a platform for the exchange of views of TB workers in the field and those in the scientific research domains, provides a further glimpse into this. From its inception in 1964 till the mid-1970s, the majority of letters, editorials, and new research articles interpreted this study's sociological insights in ways that reinscribed the central importance of individual behaviour and action, seeking to fix treatment organization and administrative lacunae with this end in mind.[49] The socio economic and political contexts of patients did not figure in the discussions, except in terms of how they affected the willingness and ability of patients to comply.[50]

Far from engaging with the wider determinants of TB, much of the writing emphasized the importance of motivation of patients before and during treatment, various means and ways of conducting retrieval actions for defaulters, and the need for behaviour change among patients.[51] Banerji wrote subsequently about the need to change the approach through which the medical sciences are employed to bring about social change, while arguing simultaneously for the need to re-orient existing notions about epidemiology of diseases and treat non-medical perspectives on health as equally important.[52] Nevertheless, the centrality of the model of default that was established by then could not be dislodged from the programme imagination.

[48] Amrith, 'In Search of a "Magic Bullet" for Tuberculosis', 121.

[49] This included studies on relationships between various actors and institutions that were part of the medical encounter—doctor–patient relationships, relationships of patients with other health personnel or paramedical staff; organizational studies of various health institutions, their functioning, accessibility, infrastructure, service delivery, administrative set ups, recording/reporting and referral system, training, supervision, communication, distances, characteristics of the population and area. B.K. Venkateshaiah, 'Technical and Operational Assessment of District Tuberculosis Programme', *NTI Newsletter* 6, no. 1 (1969): 1–7.

[50] Amrith, 'In Search of a "Magic Bullet" for Tuberculosis'.

[51] D.R. Nagpaul, 'Computer Age in TB Control', *NTI Newsletter* 5, no. 1 (1968): 11–14; M.A. Seetha and K.S. Aneja, 'Defaulter Retrieval and Cost of Defaulter Actions at Peripheral Health Institutions', *NTI Newsletter* 14, nos 3–4 (1977): 87–90; Radha Narayan and S. Pramila, 'A Model for "Motivation" of Tuberculosis Patients under the National Tuberculosis Programme', *NTI Newsletter* 9, no. 2 (1972): 20–3.

[52] Banerji, 'Tuberculosis as a Problem of Social Planning', 9–25.

WHO-IUAT Research Directions

During the 1970s–80s, the Mutual Assistance Programme of the IUAT initiated national TB programmes in Malawi, Mozambique, and Tanzania among others, and conducted several trials on chemotherapeutic regimens and their implementation in mass programmes.[53] IUAT literature of the early 1960s notes that the unrest, upheaval, and poverty in several African countries that were in the midst of struggles for independence impacted their ability to cater to the health needs of their tuberculous populations, thus necessitating close cooperation and assistance between these countries and the IUAT.[54] In spite of noting this context of sociopolitical stress with its inevitable impact on social and economic stability, several voices maintained that the problem in Africa was that in spite of the proven efficacy of domiciliary treatment in places like Madras, it was a

> formidable administrative task to translate the techniques and methods used in these carefully conducted trials to the treatment of large heterogeneous communities scattered over wide areas. Neglect in taking drugs and absconding from hospitals and clinic control are the main obstacles. These are particularly evident in patients of a low standard of mentality and education. Fox is correct in his conclusion that unrelenting pressure and education by health visitors and nurses is the only remedy.[55]

On the point of dealing with African patients having 'low mentality and education', Crofton commended the innovative practice of his colleague, Dr. Gordon, in Tanganyika, who invited to his hospital all the tribal chiefs of his district and discussed with them the struggle against TB and the part he hoped they would play in the control of

[53] C.J.L. Murray, E. DeJonghe, H.J. Chum, D.S. Nyangulu, A. Salomao, and K. Styblo, 'Cost Effectiveness of Chemotherapy for Pulmonary Tuberculosis in Three Sub-Saharan African Countries', *Lancet* 338, no. 8778 (1991): 1305–8; D.A. Enarson, 'The International Union Against Tuberculosis and Lung Disease model National Tuberculosis Programmes', *Tubercle and Lung Disease* 76, no. 2 (1995): 95–9.

[54] Etienne Bernard, 'Introductory speech at "Seminar on Tuberculosis in Africa"', 9–10 September 1960, Paris, *Bulletin of the International Journal Against Tuberculosis* 31, no. 2 (1961).

[55] Fox 1958, cited in Frederick Heaf, 'Tuberculosis in English-speaking Africa', in 'Seminar on Tuberculosis in Africa' 9–10 September 1960, Paris, *Bulletin of the International Union Against Tuberculosis* 31, no. 2 (1961): 5–10, 8.

the disease. Following this, all new patients would be admitted to the hospital for one month, during which, 'the main object was not only to start the combined chemotherapy, but chiefly propaganda to persuade the patients of the necessity of continuing their drugs even after they had become quite well'.[56] In case of patients defaulting during the remaining treatment period, the patient was to be reported to his chief, and if he 'continued to be a defaulter, to have negative urines, he was readmitted in disgrace to hospital'.[57] As health education blended with propaganda, disgrace and shame were handy tools to effect compliance; and suggestions to overcome the disease centered, here too, around the narrow confines of behaviour modification and surveillance.

The WHO and IUAT, in collaboration with the Royal Netherlands Tuberculosis Association (KNVC), set up in the mid-1960s in the Netherlands, the Tuberculosis Surveillance Research Unit (TSRU), with Karel Styblo as Director, and a mandate to study TB epidemiology relevant to control and surveillance measures.[58] During the 1970s–1980s, Styblo initiated studies in several developing countries, mainly in East Africa, on shortening the duration of the regimen and directly observing the treatment in hospital settings. Short course regimens and hospitalization were contrary to conventional methods of treatment at the time, which held this to be unsafe and cost-ineffective, leading to minimal interest in Styblo's studies at the time.[59] Ogden

---

[56] G.G. Crofton, 'Seminar on Tuberculosis in Africa', 9–10 September 1960, Paris, *Bulletin of the International Union Against Tuberculosis* 31, no. 2 (1961): 51.

[57] Crofton, 'Seminar on Tuberculosis in Africa'.

[58] Annik Rouillon, 'The Tuberculosis Surveillance Research Unit (TSRU): The First Thirty Years', *International Journal of Tuberculosis and Lung Disease* 2, no. 1 (1998): 5–9. It was at TSRU that Styblo, as Head of Scientific Activities, constructed the seminal epidemiological concept of 'risk of infection' in 1969, and subsequently in 1985 another guiding epidemiological principle labelled the 'Styblo rule'. In 1979, he became Director of Scientific Activities of the IUAT. The IUAT expanded its mandate to include other lung diseases, and changed its name to International Union Against Tuberculosis and Lung Disease (IUATLD) in 1986.

[59] Walt, Lush, and Ogden, 'International Organizations in Transfer of Infectious Diseases'; D.A. Christie and E.M. Tansey, eds, 'Short-Course Chemotherapy for Tuberculosis: The Transcript of a Witness Seminar Held by the Wellcome Trust Centre for the History of Medicine at University College London, London, on 3 February 2004', vol. 24, 144 pages (London: Wellcome Trust Centre for the History of Medicine at UCL, 2005).

et al. provide a rich account of the way that Styblo's local, context-specific studies on TB treatment were scaled up into global, universal policy under the brand name DOTS between the late 1980s and mid-1990s, by declaring chemotherapy for TB as the most cost-effective intervention available. DOTS is the acronym for 'Directly Observed Treatment, Short course', a treatment strategy in which one important component is that the patient is observed ingesting the medications by a health worker, at least for the first two months. It has four other components: ensuring political commitment; regular drug supply; diagnosis through sputum microscopy; standardized reporting system.

In the late 1980s, in the backdrop of anxiety over the 'resurgence' of TB in regions where it was thought to have been almost eliminated, the emergence of HIV/AIDS, and also WHO's declining role and influence as an international agency, the World Bank undertook a cost-effectiveness study of different health interventions as part of a Health Sector Priorities Review (HSPR), highlighting the need to refocus global attention on TB.[60] Alongside this, an independent Ad Hoc Commission on Health Research for Development emphasized that TB was the most common preventable cause of death among adults in developing countries, yet neglected.[61] Some members of the Commission were also working closely with the World Bank on a burden of disease study.[62] Within the next two years this Commission had met Styblo, done a crude cost-benefit analysis on his work in Tanzania, and found that in spite of the resource poor settings of Tanzania, 'in terms of costs per death averted and per year of life saved, chemotherapy for smear positive TB is the cheapest health intervention available in developing countries'.[63] Following this, the WHO got on board, ratified the cost-effectiveness findings through further evaluation studies, convened several meetings and workshops to chart out components and targets of a possible TB control policy strategy, and implemented a test project based on Styblo's strategy in China in 1991.

Meanwhile, a strong advocacy coalition in the WHO worked to raise the branding and marketability of the new TB control programme globally, and sought to simplify the work of the scientific

---

[60] Ogden, Walt, and Lush, 'The Politics of "Branding"'.

[61] Walt, Lush, and Ogden, 'International Organizations in Transfer of Infectious Diseases'.

[62] Ogden, Walt, and Lush, 'The Politics of "Branding"'.

[63] Ogden, Walt, and Lush, 'The Politics of "Branding"', 183.

community into 'simple, one-size-fits-all best practices which needed to be strictly adhered to'. Rapidly, a shift was made away from science and research priorities towards marketing and getting DOTS accepted in countries; this was the context in which the WHO declared TB a global emergency in 1993 and published its Framework for Effective Tuberculosis Control in 1994, thus sealing the future for the DOTS policy as the solution to TB. The WHO expert responsible for the branding and dissemination strategy held that an important mechanism of policy transfer was that 'you need to have a message that is simple enough to rally people around so that even if they don't understand it they can say that they want it'.[64] This was, however, followed by intense critiques and debates between the scientist–academic and policy–advocacy communities on the dangerous oversimplification of a complex problem, and the problematic nature of universalizing a policy of supervised therapy (branded DOTS) which might not be operationally or ethically feasible in different health service, social, and economic contexts. Despite these conflicts, however, the views of influential policy entrepreneurs and marketing and advocacy experts set the direction for the DOTS era in scientific TB policy.

Of the multiple problems associated with TB treatment in developing countries, scientific research and programme planning remained strongly focussed on the individual behavioural aspects of non-compliance and default.[65] A patient who did not adhere to the prescribed regimen was a threat to herself and the larger community, and supervised treatment was framed as the efficient response to default, right from the late 1950s. This individualization of disease led also to an individualization of health care solutions, and 'triumph over TB' was to result mainly from the TB patient being socially responsible, self-governing, adhering to medical advice, and modifying risky behaviour. This cemented forever an individualized surveillance mode of the treatment regimen, and the quest for arresting default drove the research impetus like no other issue.

Among patient-related barriers, one view on default sees patients as ignorant and unmotivated, and the range of responses include strategies for supervision, behaviour modification, increasing education and

    [64] Ogden, Walt, and Lush, 'The Politics of "Branding"', 184.
    [65] See Bayer and Wilkinson, 'Directly Observed Therapy for Tuberculosis', for a similar account of the evolution of Directly Observed Therapy in the United States.

communication activities; another sees them as requiring additional resources and support to adhere to treatment, calling for provision of incentives, transportation costs, food supplements. Health service system barriers are sought to be managed or solved through treatment re-organization, managerial and administrative restructuring, and improving the service provision in keeping with patient needs. But understanding the factors behind default in a holistic manner would need a response to wider structural factors, including socioeconomic and structural–political context related barriers which contribute to constant vulnerability of the poor.[66] This issue is left unaddressed when health and disease get abstracted from social determinants and framed as individualized occurrences.

The early phase of research on supervised therapy was also bound by a cost-effectiveness rationale, though not in the sophisticated form that took shape in the late 1980s. In a scenario of constrained state resources for providing health services, infrastructure, drugs, and given the economic costs of loss of productivity from illness, research evolved in areas that seemingly held the promise of sustained cure. Indeed, by the time of Styblo's research in Africa, the argument that supervised therapy entailed high costs could be countered forcefully by data showing higher costs of retreatment of patients who had failed or had incomplete treatment earlier.[67] The trajectory of default from supervised therapy to DOTS navigated several loops of scientific knowledge, political influence, and pragmatics of policy implementation, and worked to successfully re-structure the discussion of social determinants of TB into a matter of a five-point strategy, two of which were technical (regular drug supply and case detection through sputum microscopy) and the other three managerial (direct observation, political commitment, and standardized reporting system).

---

[66] Jaramillo, 'Encompassing Treatment with Prevention', 393–404; Paul Garner, Helen Smith, Salla Munro, and Jimmy Volmink, 'Promoting Adherence to Tuberculosis Treatment', *Bulletin of the World Health Organization* 85, no. 5 (2007): 404–6.

[67] C.J.L. Murray, A.D. Lopez, and D.T. Jamison, 'The Global Burden of Disease in 1990: Summary Results, Sensitivity Analysis and Future Directions', *Bulletin of the World Health Organization* 72, no. 3 (1994): 495–509.

ANNUAL RISK OF INFECTION, THE STYBLO RULE
AND GLOBAL ESTIMATES FOR TB

The need for 'risk of infection' information arose from concerns over the severe paucity of reliable TB morbidity and mortality information, especially in developing countries.[68] Understanding the rate of new infection was important for organizing control measures, as 'the epidemiology of tuberculosis morbidity may differ between communities with high and low risks of infection'.[69] The concept of annual risk of tuberculosis infection (ARTI, or ARI) is defined as the probability of acquiring new infection or re-infection during the course of one year, and is computed from the estimated prevalence of infection among young children tested through tuberculin skin test surveys.[70] A seminal concept developed by Styblo at the TSRU in 1969, this is thus a calculated average from an observed prevalence of infection, approximating the incidence of infection.[71] Incidence refers to new infections or cases

[68] P.V. Benjamin, 'Incidence of Tuberculosis in Economically Underdeveloped Countries and the Methods for Evaluating It' (Summary of Principal Report presented to the IUAT's XIVth International Tuberculosis Conference, held in New Delhi, January 7–11, 1957), *Indian Journal of Tuberculosis* 4, no. 2 (1957); K. Styblo and Annik Rouillon, 'Estimated Global Incidence of Smear-Positive Pulmonary Tuberculosis: Unreliability of Officially Reported Figures on Tuberculosis', *Bulletin of the International Union Against Tuberculosis* 56, nos 3–4 (1981): 118–26.

[69] Raj Narain, S.S. Nair, P. Chandrasekhar, and G.R. Rao, 'Problems Connected with the Estimation of the Incidence of Tuberculosis Infection', *Indian Journal of Tuberculosis* 13, no. 1 (1965): 5–23, 5.

[70] V.K. Chadha, S.P. Agarwal, and L.S. Chauhan, 'Annual Risk of Tuberculous Infection in Different Zones of India: A National Sample Survey, 2000–2003', in *Tuberculosis Control in India*, Directorate General of Health Services (New Delhi: Ministry of Health and Family Welfare, 2005). The process of tuberculin testing involves a representative sample of children without BCG scar being injected intradermally with a standard dose of tuberculin; the test is administered usually on the mid-inner portion of the left forearm. A rash develops at the point of injection; this is read/interpreted by trained readers after 48–96 hours and the maximum diameter of the rash is recorded. The prevalence of infection can be estimated by using three different methods to read the frequency distribution of reaction sizes that emerges.

[71] K. Styblo, J. Meijer, and I. Sutherland, 'Tuberculosis Surveillance Research Unit Report No. 1: The Transmission of Tubercle Bacilli; Its Trend in a Human

of disease; prevalence to existing infections or cases of disease. ARTI has the potential to be informative about the extent of transmission of tubercle bacilli in a community.[72]

In 1985, Styblo demonstrated another important measure, a relationship between ARTI and the incidence of smear-positive pulmonary tuberculosis, that is, new cases of pulmonary TB disease.[73] Looking at TB incidence, prevalence, and mortality data from a few countries over the period 1910–79, this paper stated that an ARTI of 1 per cent corresponded to about 50 new cases of smear-positive pulmonary TB disease per 100,000 people. While deriving this estimate, the paucity of the data at the time caused him to rely additionally on the historical epidemiological observation that deaths per year, incidence per year, and prevalence of smear-positive TB were held in the ratio 1:2:4. Further, he noted that the ratio between ARTI and prevalence of smear-positive TB was about 10, that is, each prevalent smear-positive case infects about 10 new people per year. The Styblo rule, thus, assumes a fixed mathematical relationship between the incidence of smear-positive TB, the prevalence of smear-positive TB and the ARTI (see Figure 2.1).[74]

The WHO published a technical paper applying this concept of 'annual risk of infection' to tuberculin skin test survey data from twenty-five developing countries (whereas Styblo's 1969 paper had only looked at Dutch data), and included an assessment of trends rather than current

---

Population', *Bulletin of the International Union Against Tuberculosis* 42 (August 1969): 1–104; H. Rieder, 'Annual Risk of Infection with Mycobacterium Tuberculosis', *European Respiratory Journal* 25, no. 1 (2005): 181–5.

[72] ARTI is derived from the equation, $R = 1 - (1 - P)^{1/a}$, where 'R' is ARTI, 'P' is prevalence, and 'a' is average age of the group of children tested. This is the standard approach to derive the average annual risk from prevalence of infection under the assumption of no change over calendar time. Rieder, 'Annual Risk of Infection with Mycobacterium Tuberculosis'.

[73] K. Styblo, 'The Relationship Between the Risk of Tuberculous Infection and the Risk of Developing Infectious Tuberculosis', *Bulletin of the International Union Against Tuberculosis* 60, nos 3–4 (1985): 117–19.

[74] Styblo, 'The Relationship Between the Risk of Tuberculous Infection', 117–19; F. van Leth, M.J. van der Werf, and M.W. Borgdorff, 'Prevalence of Tuberculous Infection and Incidence of Tuberculosis: A Re-assessment of the Styblo Rule', *Bulletin of the World Health Organization* 86, no. 1 (2008): 20–6.

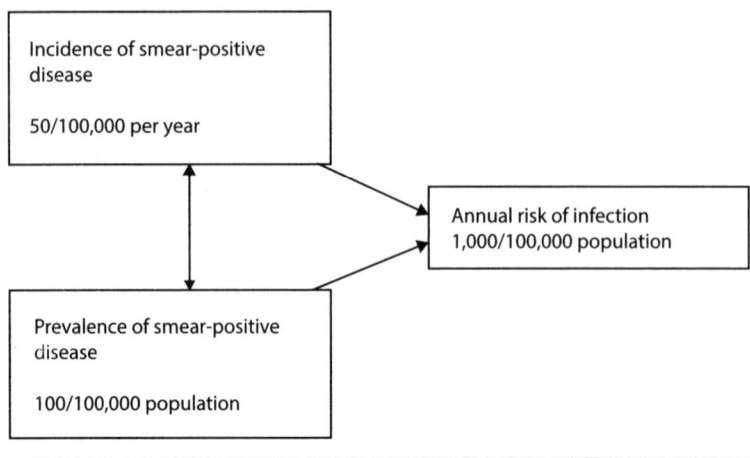

FIGURE 2.1   Mathematical relationship between ARTI, and prevalence and incidence of smear-positive TB according to the Styblo rule
Source: Figure adapted from F. van Leth, M.J. van der Werf, and M.W. Borgdorff, 'Prevalence of Tuberculous Infection and Incidence of Tuberculosis: A Re-assessment of the Styblo Rule', *Bulletin of the World Health Organization* 86, no. 1 (2008): 20–6.

status only.[75] By endorsing the concept of risk of infection (and also noting the validity of the Styblo rule), the WHO established its pivotal role in global epidemiological analyses. These two tools promptly became 'guiding rules' and a cherished part of the epidemiological canon.[76]

[75] G.M. Cauthen, A. Pio, and H.G. ten Dam, 'Annual Risk of Tuberculous Infection', Geneva, World Health Organization, 1988, unpublished document, WHO/TB/88.154.

[76] H.L. Rieder, 'Methodological Issues in the Estimation of the Tuberculosis Problem from Tuberculin Surveys', *Tubercle and Lung Disease* 76, no. 2 (1995): 114–21; C. Dye, A. Bassili, A.L. Bierrenbach, J.F. Broekmans, V.K. Chadha, P. Glaziou, P.G. Gopi, M. Hosseini, S.J. Kim, D. Manissero, I. Onozaki, H.L. Rieder, S. Scheele, F. van Leth, M. van der Werf, and B.G. Williams, 'Measuring Tuberculosis Burden, Trends, and the Impact of Control Programmes', *Lancet Infectious Diseases* 8, no. 4 (2008): 233–43.

Developing countries could now estimate new cases of smear-positive pulmonary TB from the infection rate without further investigation into other aspects.[77] Since then, the usual method to estimate the incidence of TB disease in the general population has been to take the ARTI, assessed through tuberculin skin test surveys, and apply the Styblo rule to it.[78] By the end of the 1980s, the techniques of ARI and related estimates of prevalence and incidence derived from these seminal studies were being used, along with other technical tools, in large health and disease valuation projects being rolled out by a World Bank keen on defining a new global paradigm for health care. Among these, as noted earlier, was the World Bank's HSPR, a 'series of studies on the public health significance of major clusters of disease in the developing world and on the costs and effectiveness of currently available technologies for their prevention and case management'.[79] The HSPR deployed these tools and measures to generate a scientific evidence base for a renewed paradigm of TB control, one of 'rational priority setting for health service efficiency'. This meant approaching the public health significance of these diseases from a defining frame of cost-effectiveness analysis.

In 1987 the World Bank initiated a major analytical public health initiative, the Health Sector Priorities Review. This exercise, culminating in the publication of Disease Control Priorities for Developing Countries, has documented existing knowledge about the cost-effectiveness of health interventions in developing countries. With comparable information on the cost-effectiveness of nearly 50 interventions, interest in the allocative efficiency of the health sector has increased. The broadening analytical role for cost-effectiveness laid the foundation for the health policy message in the world development report for 1993. In order to use cost-effectiveness to develop an essential package of health services, it is useful to know the burden of disease. The quantification reported here of the global and regional disease and injury burden to be addressed by the health services was thus a critical input to the World Development Report.[80]

---

[77] Styblo and Rouillon, 'Estimated Global Incidence of Smear-Positive Pulmonary Tuberculosis', 118–26.

[78] van Leth, van der Werf, and Borgdorff, 'Prevalence of Tuberculous Infection and Incidence of Tuberculosis', 20–6.

[79] Murray, Styblo, and Rouillon, 'Tuberculosis in Developing Countries', 6.

[80] Murray, Lopez, and Jamison, 'The Global Burden of Disease in 1990', 496.

As part of its exercise on estimating the burden, intervention, and cost of TB in developing countries, the HSPR employed Cauthen, Pio, and ten Dam's 1988 review data on ARI, Styblo's ARI model as well as the related Styblo rule as core techniques for assessing parameters such as TB incidence.[81] As for evidence on interventions, the independent Ad Hoc Commission on Health Research for Development had identified Styblo's short course supervised chemotherapy for TB in Tanzania as the most cost-effective intervention technique measured in terms of 'costs per death averted and per year of life saved'.[82]

By the middle of the 1990s, however, conceptual critiques of the Styblo rule started to emerge, questioning its use for estimating disease incidence on the basis of ARI. These coalesced around the point that the condition of absence of measures to control tuberculosis (such as widespread and organized chemotherapy treatments)—largely the case during Styblo's time—was no longer the case. Rieder questioned the historical assumption used in the HSPR study that for areas where information on disease incidence was lacking, the incidence could be derived by dividing prevalence by two. He noted that infection risk is 'intrinsically coupled to duration of undiagnosed, untreated transmissible tuberculosis', and intervention with chemotherapy potentially reduces the rate of transmission thus 'the average duration of infectiousness connecting prevalence and incidence becomes fundamentally changed'.[83] Also contributing to the challenge was the contemporary context of HIV infection, with HIV-associated TB either leading to increase in transmitters, or becoming symptomatic faster, leading to earlier detection and treatment with interruption of transmission.[84]

With any interruption in transmission, thus,

> the number of new tuberculous infections per prevalent smear-positive TB case will be lower than originally reported by Styblo. As a consequence, the incidence of smear-positive TB must in these circumstances

[81] Murray, Styblo, and Rouillon, 'Tuberculosis in Developing Countries', 6–24.

[82] Ogden, Walt, and Lush, 'The Politics of "Branding"', 183.

[83] Rieder, 'Methodological Issues in the Estimation of the Tuberculosis Problem', 120; also refer to Figure 2.1.

[84] Rieder, 'Methodological Issues in the Estimation of the Tuberculosis Problem', 114–21; van Leth, van der Werf, and Borgdorff, 'Prevalence of Tuberculous Infection and Incidence of Tuberculosis', 20–6.

be higher to establish an ARTI of 1 per cent. Therefore, using the Styblo rule for calculating the incidence of smear-positive TB might not be a valid approach in a situation where interventions that interrupt transmission are available.[85]

In response to Rieder's critique, there were early signs of acknowledgement from some HSPR members that re-evaluation of the data did suggest there existed 'little or no evidence that ARI and incidence are correlated in the modern era where chemotherapy is used in nearly all populations to a greater or lesser extent'.[86] Yet, the translation of this re-evaluation into the public domain remained ambiguous. 'Despite the apparent flaw in argumentation, the erroneous notion that information about the risk of infection allows the estimation of disease incidence stubbornly persist[ed], even in some prominent publications', such as the 1998 WHO *TB Handbook*.[87] The WHO Global Surveillance and Monitoring Project went on to use a modified version of the Styblo rule, among other methods, for bringing out updated estimates of global incidence of TB, while acknowledging in its report the critiques that held the potential to invalidate it.[88] Other critiques pointed out that the HSPR's disease incidence estimates, drawing from Cauthen, Pio, and ten Dam's 1988 data, ignored the latter's warning against the extrapolation of their results 'from countries covered by tuberculin surveys to countries not covered'.[89]

Apart from this, tuberculin surveys, in spite of being one of the most widely used tools for estimating prevalence of infection (in turn, ARI),

[85] van Leth, van der Werf, and Borgdorff, 'Prevalence of Tuberculous Infection and Incidence of Tuberculosis', 21.

[86] C.J.L. Murray, 'Re-examining the Annual Risk of Infection as a Monitoring Tool', unpublished draft document. Community Health Cell, Bangalore (1996).

[87] Rieder, 'Annual Risk of Infection with Mycobacterium Tuberculosis', 183; World Health Organization, *Tuberculosis Handbook* (WHO/TB/98.253) (Geneva: WHO, 1998).

[88] Christopher Dye, Suzanne Scheele, Paul Dolin, Vikram Pathania, and Mario C. Raviglione, 'Global Burden of Tuberculosis: Estimated Incidence, Prevalence, and Mortality by Country', *Journal of the American Medical Association* 282, no. 7 (1999): 677–86.

[89] Martien Borgdorff, 'Annual Risk of Tuberculous Infection: Time for an Update?' *Bulletin of the World Health Organization* 80, no. 6 (2002): 501.

have since long been fraught with problems of a technical and methodological nature. These include selection of standardized tuberculin, techniques of administration of the test, reading of the test results, digit-preference during test measurement, problem of boosting with repeated testing; difficulty of test interpretation in areas with a high presence of other kinds of mycobacteria in the environment, and also in areas with a prior history of BCG vaccination; and under-nutrition in children being tested lowering ARI estimates.[90] Combinations of these situations exist in several developing countries, posing serious challenges to interpretation of tuberculin test results, as well as making comparisons. It was finally after about a decade of critique over the Styblo rule, that WHO's Strategic and Technical Advisory Group 'endorsed the recommendation of a WHO working group to no longer use the Styblo rule to estimate the incidence of smear-positive TB' in June 2006.[91]

Developments in the late 1980s galvanized the widespread use and global legitimacy of the ARI-related tools, set in motion by a network of initiatives such as the HSPR and the Ad Hoc Commission for Health Research. While Styblo's chemotherapeutic work was taken up by the Commission for Health Research and became the DOTS strategy, his epidemiological research was taken up by the HSPR to provide burden of disease estimates for TB. In turn, these exercises also gained global legitimacy as they were able to identify pertinent

---

[90] Raj Narain et al., 'Problems Connected with the Estimation'; S.S. Nair, 'Determination of the Appropriate Index and Time for Assessing the Effectiveness of a Tuberculosis Control Programme', *Indian Journal of Tuberculosis* 24, no. 2 (1977): 58–61; Rieder, 'Annual Risk of Infection with Mycobacterium Tuberculosis', 181–5; Rieder, 'Methodological Issues in the Estimation of the Tuberculosis Problem', 114–21; van Leth, van der Werf, and Borgdorff, 'Prevalence of Tuberculous Infection and Incidence of Tuberculosis', 20–6; V.K. Chadha, H.V. Suryanarayana, M.K. Krishnamurthy, P.S. Jaganath, and A.N. Shashidhara, 'Prevalence of Undernutrition among Peri-Urban Children and Its Influence on the Estimation of Annual Risk of Tuberculosis Infection', *Indian Journal of Tuberculosis* 44, no. 2 (1997): 67–71. For the process of tuberculin testing, see footnote 70 of this chapter.

[91] van Leth, van der Werf, and Borgdorff, 'Prevalence of Tuberculous Infection and Incidence of Tuberculosis', 24; Dye et al., 'Measuring Tuberculosis Burden, Trends', 233–43.

scientific facts from the perspective of a cost-effectiveness rationale, and transform them into techno-scientific solutions for the use of TB programmes in developing countries. The unstable nature of some of these tools did not come in the way of their being taken up by influential research initiatives. The HSPR study formed the ground from which several subsequent research developments of the World Bank emerged, further fine-tuning the rationale of cost-effectiveness for translation into health policy.

It stimulated the Global Burden of Disease (GBD) study, for instance, which gave rise to yet another quantitative metric, a new composite index expressing 'burden of disease' in terms of estimated years of human life affected. This index became the 'Disability-Adjusted Life Year', DALY, incorporating diverse disease outcomes—morbidity, disability and mortality—into a single measure.[92] This aided cost-effectiveness analysis through measuring costs of specific interventions for any disease against the number of DALYs averted. These exercises combined to form the conceptual and methodological base for the *World Development Report* (*WDR*) *1993*, *Investing in Health*, which proved to be the defining feature of the turnaround that the World Bank was to then effect in global public health.[93]

However, the HSPR–GBD–WDR's attempts to calculate the 'public health significance' of major clusters of disease in the developing world through the single frame of a cost-effectiveness rationale came to be heavily critiqued, especially the tool DALYs. Paalman et al. note that 'in the original research used for the HSPR, DALYs were not used as the effectiveness/utility measure. Rather, DALYs were calculated during the HSPR review and re-working of the original research'.[94] However,

[92] C.J.L. Murray, 'Quantifying the Burden of Disease: The Technical Basis for Disability Adjusted Life Years', *Bulletin of the World Health Organization 1994* 72, no. 3 (1994): 429–45.

[93] Mohan Rao, 'Eliding History: The World Bank's Health Policies', *Disease and Medicine in India: A Historical Overview*, ed. Deepak Kumar (Delhi: Tulika Publishers, 2001); World Bank, *World Development Report 1993: Investing in Health* (New York: Oxford University Press, 1993).

[94] Maria Paalman, Henk Bekedam, Laura Hawken, and David Nyheim, 'A Critical Review of Priority Setting in the Health Sector: The Methodology of the 1993 World Development Report', *Health Policy and Planning* 13, no. 1 (1998): 13–31.

the crude version of the cost-effectiveness calculation for the HSPR TB study—comparison between standard chemotherapy and short-course chemotherapy in terms of 'cost per case treated; cost per case cured at eighteen months; cost per death averted (estimates of 'death averted' were derived, again, from estimates linked to the Styblo rule); cost per death averted including one round of transmission'—indicated that the core principles of future DALY calculations had already been integrated into this study.

There was a wide consensus, among the different conceptual and technical critiques that emerged, that 'burden of disease' was defined in normative terms, excluding socioeconomic factors which have a significant influence on the actual burden experienced by an individual, family or community.[95] The measure was criticized for having moved the idea of health back from 'well-being' to a greater disease orientation and focus on health *care*, with the entire exercise aiming to 'justify technological interventions and undermine socio-political dimensions of health'.[96] Thus, not only the evidence base but even the scientific technologies deployed for TB programmes during these decades purged any possibility of a socioeconomic or a political economy perspective on TB disease and treatment; the question of TB as a social disease remained bounded by cost-effective biomedical technical interventions for infectious disease control.

★ ★ ★

This chapter has attempted to lay out some of the processes behind the re-casting of the problem of poverty in TB through a techno-scientific framing. It has discussed the ways this has been articulated through an individualized, behavioural framing of the structure of cure, as well as a cost-effectiveness framing of the attainment of health. It has shown the ways this framing has attained legitimacy, and been translated from

---

[95] Paalman et al., 'A Critical Review of Priority Setting in the Health Sector', 13–31; Anand Sudhir and Kara Hanson, 'Disability-Adjusted Life Years: A Critical Review', *Journal of Health Economics* 16, no. 6 (1997): 685–702.

[96] Ritu Priya, 'Disability Adjusted Life Years as a Tool for Public Health Policy: A Critical Assessment', in *Public Health and the Poverty of Reforms: The South Asian Predicament*, ed. Imrana Qadeer, Kasturi Sen, and K.R. Nayar (New Delhi: Sage Publications, 2001).

local to global contexts under the aegis of influential scientific and policy-making initiatives.

This discussion has shown that given the unstable and dynamic nature of scientific judgements, such judgements tend to be 'artefacts of the chosen methodologies rather than [...] representations of reality'.[97] The chapter has tried to elaborate the manners in which scientific judgements get validated from being 'backed up by an organizational infrastructure', and the ways this tends to 'emphasize those aspects of risk that its scientific bureaucracy has the tools to measure'.[98]

For the most part, this chapter has tried to illustrate that poverty appearing to hold sway over the policy imagination as a factor impacting TB is merely a mirage; the development of scientific tools and policy in TB has consistently ruptured any possibility of engaging with the structural, political, and discursive elements of poverty in TB.

---

[97] Hornstein, 'Reclaiming Environmental Law', 573.
[98] Hornstein, 'Reclaiming Environmental Law', 573–5.

3   'Surveillance for Equity'?

Poverty, Inequality, and the Anti-politics of Family Planning

Rebecca Williams

During the 1970s, health inequalities became a key issue for international health experts.[1] As part of this increasing concern with inequality, many international health practitioners began to promote the concept of 'equity' in health. Since the 1970s, advocates of the concept have offered various definitions of, and approaches to, health 'equity'.[2] Most commonly, proponents of health 'equity' define

[1] Nivaldo Linares-Pérez and Oliva López-Arrelano, 'Health Equity: Conceptual Models, Essential Aspects, and the Perspective of Collective Health', *Social Medicine* 3, no. 3 (2008): 194–206; Davidson R. Gwatkin, 'Health Inequalities and the Health of the Poor: What Do We Know? What Can We Do?' *Bulletin of the World Health Organization* 78, no. 1 (2000): 4.

[2] Linares-Pérez and López-Arrelano, 'Health Equity', 194–206; Margaret Whitehead, 'The Concepts and Principles of Equity and Health', (WHO Regional Office for Europe Working Paper, 1990), 3; Paula Braveman, 'Health Disparities and Health Equity: Concepts and Measurement', *Annual Review of Public Health* 27 (2006), 167–94; P. Braveman and S. Gruskin, 'Defining Equity in Health', *Journal of Epidemiology and Community Health* 57, no. 4 (2003): 257; A.J. Culyer and Adam Wagstaff, 'Equity and Equality in Health and Health Care', *Journal of Health Economics* 12, no. 4 (1993): 431–57; James A. Macinko and Barbara Starfield, 'Annotated Bibliography on Equity in Health, 1980–2001',

inequities as those inequalities in health which they deem unfair or unjust.[3] Health 'equity' projects attempt to rectify these inequalities through technical interventions which typically are narrowly focused on health outcomes, and centrally concerned with the health of the poor.[4]

One influential proponent of 'equity' in health was Carl E. Taylor, Professor of International Health at the Johns Hopkins School of Public Health (JHSPH). This chapter explains how, from the mid-1970s, Taylor advocated 'equity' in health as a means to improve the health of poor populations in 'developing' nations.[5] Importantly, Taylor upheld 'systematic monitoring' of the population—rather than political, social, and structural change—as the solution to inequalities in health.[6] Specifically, Taylor advocated the implementation of epidemiological methods of surveillance as a route to health 'equity'.[7] In approach which he called 'surveillance for equity', Taylor argued that, through systematic monitoring, community health workers could improve the health of poor populations in the absence of socioeconomic equality, and at minimal cost.[8]

International Journal for Equity in Health 1, no. 1 (2002), http://www.equity-healthj.com/content/1/1/1 (accessed 12 April 2012).

[3] Whitehead, *Concepts and Principles of Equity and Health*, 5; Timothy Evans et al., 'Introduction' in *Challenging Inequities in Health: From Ethics to Action*, ed. Timothy Evans, Margaret Whitehead, Finn Diderichsen, Abbas Bhuiya, and Meg Wirth (Oxford: Oxford University Press, 2001), 4; Andrew Ward, Pamela Jo Johnson, and Mollie O'Brien, 'The Normative Dimensions of Health Disparities', *Journal of Health Disparities Research and Practice* 6, no. 1 (2013): 47.

[4] Fabienne Peter, '*Health Equity and Social Justice*', in *Public Health, Ethics, and Equity*, ed. Sudhir Anand, Fabienne Peter, and Amartya Sen (Oxford: Oxford University Press, 2004), 94; Davidson R. Gwatkin, 'The Need for Equity-oriented Health Sector Reforms', *International Journal of Epidemiology* 30, no. 4 (2001): 720–3.

[5] Carl E. Taylor, 'Surveillance for Equity in Primary Health Care: Policy Implications from International Experience', *International Journal of Epidemiology* 21, no. 6 (1992): 1043–9.

[6] Taylor, 'Surveillance for Equity', 1043.

[7] Taylor, 'Surveillance for Equity', 1043.

[8] Taylor, 'Surveillance for Equity', 1043.

In this chapter, I argue that 'surveillance for equity' is what James Ferguson terms an 'anti-political' project.[9] By focusing narrowly on health inputs and outputs, Taylor's model of health equity 'systematically erased' the political and structural causes of poverty, and replaced them with technical causes, requiring technical solutions.[10] In short, 'surveillance for equity' isolated the bodies of the poor from their social, economic, and political contexts, and transformed them into objects of medical knowledge and management. In its attempt to provide an immediate solution to the diseases of poverty, Taylor's health 'equity' project sidelined the problem of poverty per se in favour of a limited focus on specific health indices.

To understand the genesis of Taylor's approach to 'equity' in health, this chapter will first turn to the system of surveillance which Taylor promoted as a means to rectify health inequalities. I trace this system of surveillance to the Khanna study, a well-known population control experiment which Taylor and his colleagues from the Harvard School of Public Health (HSPH) conducted in the Ludhiana District of Punjab during the 1950s.[11] The Khanna study took a system of continuous population monitoring, derived from communicable disease study, and applied it to population control. Subsequently, Taylor and his colleagues adapted this epidemiological system of surveillance to a broader set of family planning, nutrition, and community health interventions. By tracing the adaptation of this mode of surveillance from the Khanna study to health 'equity', this chapter shows that 'surveillance for equity' was not a new approach to health care, but an existing set of technical interventions repackaged in the language of social justice.

THE KHANNA STUDY AND THE FAMILY FOLDER SYSTEM: A GENEALOGY OF SURVEILLANCE

In the post-war period, a number of government and non-government organizations, private foundations, and academic institutions became

---

[9] James Ferguson, *The Anti-Politics Machine: Development, Depoliticization, and Bureaucratic Power in Lesotho* (Cambridge: Cambridge University Press, 1990), 14–15, 66.

[10] Ferguson, *Anti-Politics Machine*, 66.

[11] John B. Wyon and John E. Gordon, *The Khanna Study: Population Problems in the Rural Punjab* (Cambridge: Harvard University Press, 1971).

vitally concerned about population growth across the so-called under-developed or third world.[12] Carl Taylor, John Wyon, and John Gordon, epidemiologists from the HSPH, were among those concerned with third world 'overpopulation'. Taylor, Wyon, and Gordon argued that 'population pressure' was a 'social malady' that could lead to famine, poverty, and civil unrest.[13] In particular, they argued that population growth presented a public health problem.[14] In a theory that they called the 'epidemiology of population', Gordon, Taylor, and Wyon argued that India, with a young population and 'high natural potential of growth', posed 'the most serious economic and public health problem' to the entire world.[15]

During the early 1950s, Gordon, Taylor, and Wyon designed a research project to study 'population dynamics' in a series of villages surrounding the market town of Khanna, in the Ludhiana District of Punjab.[16] In addition to collecting information on the reproductive processes and sexual practices of the populations of these villages, the Khanna study aimed to test the effects of five 'simple methods' of

[12] See, for example, John Sharpless, 'Population Science, Private Foundations, and Development Aid: The Transformation of Demographic Knowledge in the United States, 1945–1965', in *International Development and the Social Sciences*, ed. Frederick Cooper and Randall Packard (Berkeley: University of California Press, 1997), 176–200; John Sharpless, 'World Population Growth, Family Planning and American Foreign Policy', *Journal of Policy History*, 7, no. 1 (1995): 72–102; Eric B. Ross, *The Malthus Factor: Poverty, Politics and Population in Capitalism Development* (London; New York: Zed Books, 1998); Matthew Connelly, *Fatal Misconception: The Struggle to Control World Population* (Cambridge: Belknap Press, 2008); Matthew Connelly, 'Population Control Is History: New Perspectives on the Campaign to Limit Population Growth', *Comparative Studies in Society and History* 45, no. 1 (2003): 122–47.

[13] Wyon and Gordon, *Khanna Study*, 18.

[14] John E. Gordon and Theodore H. Ingalls, 'Public Health as Demographic Influence', *American Journal of the Medical Sciences* 227, no. 3 (1954): 326–57.

[15] John Gordon, John Wyon, and Carl Taylor, 'India-Harvard-Ludhiana Study of Population Control: Part 1: Public Health Aspects', Rockefeller Foundation Records (hereafter cited as Rockefeller Foundation Records) Rockefeller Archive Center, Record Group 1.2, Series 200, Box 45, Folder 373, 5.

[16] Wyon and Gordon, *Khanna Study*, 19.

contraception on the birth rate.[17] By the end of the Khanna study, its architects and funders alike considered the project a 'glorious failure'.[18] The study had not achieved the rates of 'acceptance' of family planning methods required to affect the birth rate.[19] Nevertheless, Government of India officials cited the Khanna study as evidence that the Indian people were ready and willing to limit the size of their families.[20]

Subsequently, the Khanna study gained infamy thanks to Mahmood Mamdani's *The Myth of Population Control* (1972). After spending several months interviewing villagers in one of the Khanna study experimental villages, Mamdani argued that the Khanna Study staff had failed to understand that most families in Manupur *needed* large families for economic and physical security.[21] Furthermore, Mamdani argued that the study had been plagued by the basic understanding of the problem, shared by the study's architects and Indian staff alike, of population as a disease to be treated with the techniques of an epidemiologist.[22]

Indeed, warning of a 'pandemic of births', Gordon, Taylor, and Wyon had designed the Khanna study as an epidemiological investigation to address the 'disease' of 'overpopulation'.[23] Having framed population growth as a public health problem, the Harvard epidemiologists adapted procedures of mapping, census taking, and preparation of record forms 'with the family as the unit of observation' from communicable disease study.[24] Gordon, Taylor, and Wyon implemented a

[17] Wyon and Gordon, *Khanna Study*, 13–15.

[18] Clyde V. Kiser, ed., *Research in Family Planning* (Princeton: Princeton University Press, 1962), 614.

[19] 'Manupur', undated: 1, Rockefeller Foundation Records, Record Group 1.2, Series 200, Box 45, Folder 371.

[20] Government of India, *Report of the Family Planning Programme Evaluation and Planning Committee*, Part I: *'Problem'* (n.d., c. 1966): 73, National Documentation Centre, National Institute of Health and Family Welfare, New Delhi (hereafter NDC), 204/83 F.

[21] Mahmood Mamdani, *The Myth of Population Control: Family, Caste and Class in an Indian Village* (New York: Monthly Review, 1972), 32.

[22] Mamdani, *Myth of Population Control*, 37.

[23] Gordon, Wyon, and Taylor, 'India–Harvard–Ludhiana Study of Population Control: Part 2: The Epidemiology of Population', 41.

[24] John Gordon, 'Progress Report—India–Harvard–Ludhiana Population Study December 1, 1955', 5, Rockefeller Foundation Records, Record Group 1.2, Series 200, Box 46, Folder 375.

system of demographic surveillance, centred upon a 'family folder', to monitor and intervene upon the Khanna study population.

Gordon had utilized such a system of surveillance to map the epidemiology of scarlet fever in 1930s Romania.[25] In that study, field staff began by mapping the location of every building in the study villages, and assigning each home with a number. They then made a census list of the inhabitants of each home by house-to-house visits, and assigned each person a code—based on their house number—a letter to indicate his relationship in the family—for example, a father was A, a mother was B, a child C; thus, a father in house number 13 would be 13A. This code, along with the person's name, appeared on all other records relating to that individual. Next, field staff prepared a history folder for each home. This folder listed the members of the family, their relationship, sex, date of birth, occupation, and whether they had ever been immunized against scarlet fever. Field workers fastened individual history sheets for each family member to the folder, to help keep a running record of visits and record any illness. A public health nurse resident in the village visited each family at least once a month. She kept a record of her findings on daily work sheets, which she sent to the base office along with cultures from suspected cases of infection. Office staff recorded all data collected for any one person on the individual record sheet for that person in their family folder.[26]

The method of surveillance which Gordon, Taylor, and Wyon employed in the Khanna Study was a modified version of this family folder-based system, but tailored to the collection of information about reproduction. For example, field workers used baseline census data to identify all couples who they deemed 'eligible' for family planning advice by identifying every woman between the ages of 15 and 44 from the family record.[27] Field staff also collected data on the 'nature and behaviour of menstruation' among women in the study villages, to enable them to time their monthly visits to coincide with the expected

[25] John E. Gordon, 'The Epidemiology of Scarlet Fever', *American Journal of Epidemiology* 38, no. 1 (1943): 27–98.

[26] Gordon, 'The Epidemiology of Scarlet Fever', 30–1.

[27] John Gordon, 'Exploratory Investigation—Population Dynamics—Chakohi Village, Punjab, India June 1954–March 1955': 92, Rockefeller Foundation Records, Record Group 1.2, Series 200, Box 45, Folder 375.

date of menstruation.[28] Once field workers knew when to expect menstruation, they could make a 'diagnosis of pregnancy' on the basis of missed period.[29]

Gordon, Taylor, and Wyon replicated and disseminated this system of surveillance—based on maps, censuses, and regular house-to-house visitation—throughout their research and teaching careers. Research studies and community health projects in a number of locations worldwide, from Haiti to Bangladesh, implemented systems of population surveillance which continued the 'tradition of home visitation' which Wyon, Taylor, and Gordon utilized in the Khanna study.[30] The Department of Community Medicine of the Ludhiana Christian Medicine College (CMC)—which collaborated with the HSPH in the Khanna study—has, since the 1970s, used the family folder system to continuously monitor the populations of its field practice areas. Community health workers use this system of surveillance to identify and make specific interventions into 'high risk' or 'priority needs' households.[31] Several government and non-government institutions across India have, in turn, adopted the family folder system of the Ludhiana CMC for community health work.[32]

[28] Gordon, 'Exploratory Investigation', 74–9.

[29] Gordon, 'Exploratory Investigation', 82.

[30] K.M. Aziz and W. Henry Mosley, 'The History, Methodology, and Main Findings and the Matlab Project in Bangladesh', in *Prospective Community Studies in Developing Countries*, ed. Monika Das Gupta, Peter Aaby, Michel Garenne, and Gilles Pison (Oxford: Clarendon Press, 1998); Henry B. Perry, 'Primary Health Care in Bangladesh: Challenges, Approaches, and Results', in *Community-Based Health Care: Lessons from Bangladesh to Boston*, ed. John Rohde and John Wyon (Boston: Management Sciences for Health, 2002), 38–9; Gretchen Berggren, Warren Berggren, Henri Menager, and Eddy Genece, 'Longitudinal Community Health Research for Equity and Accountability in Primary Health Care in Haiti', in *Prospective Community Studies*, ed. Monika Das Gupta, Peter Aaby, Michel Garenne, and Gilles Pison (New York: Clarendon Press, 1997), 157–8.

[31] Betty Cowan, 'Community Health Department Programme, Christian Medical College and Brown Memorial Hospital, Ludhiana, Punjab, India', July 1975: 2, Wellcome Library, Archives and Manuscripts Collection, Stanley George Brown Collection, WTI/SGB/J.3.

[32] Department of Community Health and Social and Preventive Medicine, *Annual Report 1984*: 2; Department of Community Health and Social and

Most important for the purposes of this chapter were the uses to which Taylor put this family folder-based system of surveillance after the Khanna study. During the mid-1960s to mid-1970s, Taylor adapted the field procedures of the Khanna study to a field study at Narangwal, also in the Ludhiana District of Punjab.[33] The Narangwal study consisted of two interconnected research projects: a nutrition study and a population study. Through the nutrition study, Taylor and his colleagues from the Department of International Health at the JHSPH set out to demonstrate that a clear relationship existed between nutrition and infection. Specifically, the Narangwal nutrition study aimed to test the hypothesis that villages in which community health workers provided both nutritional supplements and medical care would have better health outcomes than villages in which health workers supplied either medical or nutritional services alone.[34]

Meanwhile, the Narangwal population study attempted to show that an integrated programme of maternal and child health care and family planning services would be more effective at reducing the birth rate than the so-called 'vertical' interventions—such as the Khanna study—which offered family planning services alone. Underpinning this approach was what Taylor called the 'child survival hypothesis'. That is, the idea that couples would be more likely to limit the size

---

Preventive Medicine, *Annual Report 1985*: 2, Ludhiana CMC, Department of Community Health Library (hereafter Department of Community Health Library); J.P. Majra and Acharya Das, 'Impact of Family Folder System on the Health Status of the Community', *Internet Journal of Healthcare Administration* 6, no. 2 (2009), http://www.ispub.com/journal/the-internet-journal-of-healthcare-administration/volume-6-number-2/impact-of-family-folder-system-on-the-health-status-of-the-community.html (accessed 5 February 2012); website of St. Stephen's Hospital, New Delhi, http://www.sshchd.org/content.php?id=2andsid=13andsid2=17andmid=13andlevel=2 (accessed 28 February 2012).

[33] Carl E. Taylor, 'Origins of Longitudinal Community-Based Studies', 19–27; Carl E. Taylor and Cecile De Sweemer, 'Lessons from Narangwal about Primary Health Care, Family Planning, and Nutrition', *Prospective Community Studies* (1997): 104.

[34] Arnfried A. Kielmann et al., 'Child and Maternal Health Services in Rural India: The Narangwal Experiment', *Integrated Nutrition and Health Care*, vol. 1 (Baltimore; London: Johns Hopkins University Press, 1983).

of their families if they could be sure that their existing children were likely to survive until adulthood.[35] Parallel to these two main academic research goals, the Narangwal study also developed a package of integrated health care which combined family planning, women's services, child care, and nutrition. The Narangwal study team aimed to demonstrate how such a package could be rolled out, across India and beyond, at minimal cost.[36]

Lastly, from the mid-1970s, Taylor argued that this system of surveillance, centred upon family folders and house-to-house visitation, could enable public health practitioners to address a rapidly-growing concern among international health experts: 'equity' in health. What is 'equity' in health, and how did Taylor believe that surveillance would deliver it?

'SURVEILLANCE FOR EQUITY'?

Taylor conceptualized 'equity' in health as a means of delivering equality of health outcomes in the absence of socioeconomic justice. In other words, health 'equity' was a means to mitigate the physical manifestations of poverty—in terms of increased ill-health—without addressing the root causes of poverty. Taylor's approach to 'equity' in health is an explicit example of what James Ferguson terms the 'anti-politics' of development projects. That is, the process whereby the political and structural causes of poverty are 'systematically erased and replaced with technical ones' requiring technical solutions.[37] While sidelining the structural causes of poverty, the 'anti-politics machine' of development simultaneously expands and entrenches bureaucratic state power.[38]

In an effort to provide immediate, practical solutions to the physical symptoms of poverty, Taylor and his colleagues produced a blueprint for a system of health intervention to improve the health of the poor

[35] Carl E. Taylor et al., *Integrated Family Planning and Health Care*, vol. 2, in *Child and Maternal Health Services in Rural India: The Narangwal Experiment* (Baltimore; London: Johns Hopkins University Press, 1983), 7.

[36] Taylor et al., *Integrated Family Planning and Health Care*, 13.

[37] Ferguson, *Anti-Politics Machine*, 14–15, 66.

[38] Ferguson, *Anti-Politics Machine*, 251–77.

without improving their socioeconomic status. In so doing, health 'equity' isolated the bodies of the poor from their social, economic, and political contexts, and transformed them into objects of medical knowledge and management. In addition, Taylor promoted 'surveillance for equity' as a way to improve health at low cost; 'equity' in health outcomes was also compatible with efficiency of health care provision. Thus, 'surveillance for equity' required economic redistribution and social change neither within nor across nations. In place of political and structural equality, Taylor's model provided the poor with continuous monitoring.

Taylor began to advocate 'equity' in health during the mid-1970s. He crystallized his ideas within a 1992 article entitled, 'Surveillance for Equity in Primary Health Care: Policy Implications from International Experience'.[39] In this article, Taylor outlined what he described as a 'management paradigm' for improving health using epidemiological methods of surveillance. In summary, Taylor argued that health could be controlled by 'identifying subgroups among whom disease and risk are concentrated, defining priority problems, determining underlying causation and implementing corrective interventions'.[40] Taylor suggested that international organizations should make the implementation of such a health care system a condition of their aid.[41]

In his 'Surveillance for Equity' article, Taylor offered an 'operational definition' of 'equity' as the 'distribution of benefits according to demonstrable need rather than on the basis of political or socioeconomic privilege'.[42] Taylor made an important distinction between the concept of 'equity' in health and alternative notions of equality. He explained that his approach to health 'equity' differed from others which aimed to 'promote fairness' in that it focused on 'disparities in health status', rather than upon inequalities in provision of, or access to, health services.[43] In short, Taylor argued that his model of 'surveillance for equity' could produce equitable health *outputs*—expressed as a reduced disparity between different social groups in specific,

---

[39] Taylor, 'Surveillance for Equity', 1043–9.
[40] Taylor, 'Surveillance for Equity', 1043.
[41] Taylor, 'Surveillance for Equity', 1048.
[42] Taylor, 'Surveillance for Equity', 1043.
[43] Taylor, 'Surveillance for Equity', 1043.

measurable health indices—rather than equality of health care *inputs*. Distribution of, and access to, health care were not at issue in the 'surveillance for equity' model. Nor was the broader question of socioeconomic inequality—with its known effects on health—under scrutiny. At stake was the distribution of health itself. I will return to this important distinction shortly. However, in order to further understand the concept of 'surveillance for equity', this chapter will first outline how Taylor proposed this system should operate in practice.

Within Taylor's model, the key to achieving health 'equity' was the establishment of surveillance, which Taylor defined as 'an information system to identify emerging problems promptly so as to respond rapidly with appropriate action'.[44] The system of surveillance which Taylor proposed to use to deliver health 'equity' was a modified version of the epidemiologic methodology which—as I described earlier in this chapter—the Khanna Study had employed. In the same way as the Khanna Study had continuously surveyed its study villages in an attempt to alter birth rates, 'surveillance for equity' consisted of the 'systematic monitoring' of a district population. The first step towards establishing this system was the registration of all families in the district.[45] Next, health workers would work with 'community leaders' to identify 'priority problems' and decide how to allocate resources most efficiently.[46] As part of this process, health workers would also identify what Taylor called 'equity indicators'. That is, specific health indices—for example, the distribution of births and deaths, and child growth—in which disparities between socioeconomic groups were visible. By identifying these 'equity indicators', health workers could concentrate their efforts among 'high risk subgroups' whose health measured poorest against these indices.[47]

Once health workers had registered all families, identified priorities and established 'equity' indices, they would continue to monitor the community by visiting each home at least once a year. Community health workers would visit homes in poorer areas, in which they had identified 'high risk' subpopulations, more frequently. For example, a

[44] Taylor, 'Surveillance for Equity', 1043.
[45] Taylor, 'Surveillance for Equity', 1043.
[46] Taylor, 'Surveillance for Equity', 1043–5.
[47] Taylor, 'Surveillance for Equity', 1044.

population subgroup with high death rates and comparatively poor rates of child growth would be a prime target for more intensive surveillance. By paying particular attention to these 'high risk' homes, health workers could identify and treat illnesses—or potential ill-nesses—at an early stage, using 'simple and low-cost interventions'.[48] In this way, community health programmes could level disparities in 'equity indicators' between social groups, and maximize health outputs without adding extra services. Thus, Taylor argued that 'equity and efficiency' could be 'complementary'.[49]

'Surveillance for equity' therefore meant continuous, intensive tar-geting of the bodies of the poor to equalize specific health indices without additional health care inputs. As the most socioeconomically disadvantaged people were transformed into 'high-risk' medical sub-groups, their poor health was converted from a symptom of poverty to a set of indices to be improved through rational medical management. Indeed, Taylor considered poor households not only to be the sites of highest disease concentration, he also considered these households to be the least able to manage their own health. Taylor explicitly stated that health workers should provide the most intensive surveillance for 'the poor and least educated'. By contrast, Taylor suggested that 'educated' families could be taught to conduct much of their own monitoring—for example, by weighing children—and initiate their own care.[50] Within the 'surveillance for equity' model, therefore, the poor are framed as singularly incapable of managing their own health, and in need of expert oversight.

Taylor cited the example of the Narangwal Study as evidence that his approach to 'equity' could work in practice.[51] Taylor and his colleagues discussed the Narangwal study in terms of health 'equity' in the final monographs of the Narangwal population and nutrition projects.[52] In these works—as in Taylor's 'surveillance for equity' article—the tar-geting of the poor and the rational management of health resources

[48] Taylor, 'Surveillance for Equity', 1044.

[49] Taylor, 'Surveillance for Equity', 1044.

[50] Taylor, 'Surveillance for Equity', 1043–4.

[51] Taylor, 'Surveillance for Equity', 1046–7.

[52] Arnfried A. Kielmann et al., *Integrated Nutrition and Health Care*, vol. 1 (1983), 1; Taylor et al., *Child and Maternal Health Services in Rural India*, 2.

were central to the concept of 'equity'. Indeed, the authors of the Narangwal Study volumes defined 'equity' as 'measuring the provision of services to target groups that are underserved or disadvantaged'.[53] The Narangwal study team claimed that through 'continued intensive surveillance and preferential care for the underserved population (low-caste villagers)', the Narangwal study had reduced disparities in health between socioeconomic groups.[54]

In the final monograph of the Narangwal population project, the study team offered the use of family planning as an illustration of how the Narangwal study had created 'equity'. Taylor and his colleagues pointed out that, at the beginning of the project, 21 per cent of high-caste couples and 13 per cent of low-caste couples had ever used a 'modern' family planning method. By the end of the study, this gap was 'equalized': 46 per cent of couples from both caste groups practiced family planning at some point during the project.[55] The authors of the Narangwal study claimed that these statistics offered 'clear-cut' evidence that 'if adequate coverage of health services, education, and nutrition can be provided—the poor seem to be as willing to limit their fertility as the rich'.[56] This, Taylor and his colleagues claimed, demonstrated how the project had 'eliminated disparities' through surveillance and technical interventions.[57]

Yet, does this truly constitute 'equity'? One way to begin to answer this question is through the following comparison. During my research, I was struck by the similarity between the 'surveillance for equity' model which Taylor and his colleagues implemented at Narangwal, and the 'priority homes' and 'high risk families' approach of the Department of Community Health of the Ludhiana CMC. Similarly to the Narangwal study, staff at the Ludhiana CMC described their methodology as one 'through which everybody in the community is reached and services provided according to present known and antici-pated health needs'.[58] Both the Narangwal study and the Ludhiana

[53] Taylor et al., *Integrated Family Planning and Health Care*, 47.

[54] Taylor et al., *Integrated Family Planning and Health Care*, 34.

[55] Taylor et al., *Integrated Family Planning and Health Care*, 49–50.

[56] Taylor et al., *Integrated Family Planning and Health Care*, 47.

[57] Taylor et al., *Integrated Family Planning and Health Care*, 49.

[58] *Community Health Department, Ninth Annual Report 1980*, Department of Community Health Library, CMC Ludhiana, 1980: 1.

CMC employed a system of surveillance, centred upon regular house-to-house visits. In each project, health workers used the family folder system to identify, monitor, and target high-priority health problems among high-risk subgroups in their respective populations. The similarity in their approaches was not a coincidence as the staff at the Ludhiana CMC was familiar with the family folder systems employed in both the Khanna and Narangwal studies.[59] Notably, however, staff at the Ludhiana CMC did not attach the term 'equity' to their work. The key terms which the staff of the Ludhiana CMC applied to their work were 'priority' and 'risk'.

What, then, is the difference between the 'priority homes' approach of the Ludhiana CMC and 'surveillance for equity'? The main difference is simple, but significant: language. In fact, highly revealing differences, shifts, and changes in language exist not only between these two family folder-centred health projects, but within Taylor's own body of work. This chapter will now turn to these linguistic shifts.

LANGUAGE SHIFT

'Surveillance for Equity' was not a new approach to health care, but an existing set of technical interventions translated into the language of social justice. Indeed, a careful chronological reading of Taylor's works reveals that 'equity' was a term which he applied only retrospectively to the methodology and outcomes of the Narangwal Study. Taylor began to discuss health 'equity' in his published works from the mid-1970s onwards. He and his colleagues dedicated considerable space to the

[59] Particularly Dr Harbans Kaur Dhillon who, as head of the Department of Community Health, introduced the family folder system to the health centres of the Ludhiana CMC in 1972. Dhillon served on the Indian Advisory Committee to the Khanna Study during 1953–60. In 1970, Taylor identified Dhillon as a particular supporter for the Narangwal Study. *Department of Community Health and Social and Preventive Medicine, Annual Report 1982*: 1, Department of Community Health Library; 'Minutes of the 4th meeting of the Advisory Committee of the India–Harvard–Ludhiana Population Study held in Khanna at 9:30 a.m. on April 4th 1956', RAC, RF, Record Group 1.2, Series 200, Box 45, Folder 370; Narangwal RHRC, *Integration of Health and Family Planning in Village Sub-Centres: Report on the Fifth Narangwal Conference, November 1970* (Narangwal: RHRC, 1970): 1.

concept in the definitive monographs of the Narangwal projects, which were published in 1983.[60] However, the term did not appear in the reports or conference proceedings of the Narangwal Study which were produced during the research period itself—1967–74. These earlier reports described several ways in which the Narangwal study's system of surveillance was crucial to the performance of the study. Yet, study staff did not connect these surveillance practices to 'equity' in health until later. To illustrate this shift, this chapter will turn to two examples of the changing ways in which Narangwal study staff conceptualized the use of surveillance: family planning and the monitoring of 'risk'.

Family Planning

The annual reports of the Narangwal population project explained how the project used data, collected through routine surveillance, to develop a practice of 'systematic family planning'.[61] That is, a practice of offering family planning services not only to those who sought them, but to all couples whom community health workers identified as 'eligible'. The study defined 'eligible' couples as those where the wife was in the 'reproductive age group' of 15–49. [62] Family health workers were instructed to offer family planning advice to 'non-users' of contraceptives on a number of 'obligatory occasions' or 'entry points'. These occasions included: during a routine fertility survey; upon confirmation of pregnancy; during a post-abortion examination visit and during a post-partum examination visit.[63] The annual reports of the Narangwal study explain that such a 'systematic' approach allowed the project staff 'to contact a larger proportion of the eligible population' with family planning advice than they would otherwise be able to contact. In addition, this approach allowed health workers to identify those women whom they considered 'highest priority in their eligibility for family

[60] Taylor et al., *Integrated Family Planning and Health Care*, 32–4, 47–51, 191–213; Kielmann et al., *Integrated Nutrition and Health Care*, vol. 1, 1983, 63–5, 85–7.

[61] Narangwal Rural Health Research Center (RHRC), *Population Project Annual Report to the ICMR 1971–1972* (1972), 89, NDC.

[62] Taylor et al., *Integrated Family Planning and Health Care*, 158.

[63] Narangwal RHRC, *Population Project Annual Report 1971–1972*, 89.

planning'.[64] Crucially, the aim of the Narangwal population project was to reduce the birth rate of the study population by identifying, monitoring, and targeting 'eligible' women. Taylor and his colleagues implemented their system of surveillance to achieve this goal.

The contract between Johns Hopkins and the United States Agency for International Development, which funded the Narangwal population project, did not speak of achieving 'equity'. Rather, it identified the objectives of the Narangwal population project as strengthening 'the factual and analytical base for AID policy decisions with respect to health related systems for delivering family planning services'. In particular, the researchers and funders of the Narangwal study hoped 'to enhance the effectiveness of AID fertility control assistance to India'.[65] In other words, the main objective of the Narangwal study was to make population control programmes more effective at lowering birth rates. As I explained earlier in this chapter, the main way in which Taylor and his colleagues hoped to influence family planning policy was by testing the 'child survival hypothesis'. Their principal task was to assess whether the combination of family planning services with health care would increase 'acceptance' of family planning methods.[66] The Narangwal study team developed their field methodology with the aim of demonstrating that the integration of health and family planning services was 'feasible' and an effective route to population control.[67]

[64] Narangwal RHRC, *Population Project Annual Report to the ICMR 1971–1972*, 88; Narangwal RHRC, *Rural Health Services and Family Planning Utilization: Annual Report of Population Research 1972–73*, 23, Department of Community Health Library, CMC Ludhiana.

[65] 'Contract between the United States of America and the Johns Hopkins University, PIO/T298-53-6296121', 30 June 1969: B-1, Johns Hopkins University, Alan Mason Chesney Medical Archives, Carl Taylor Collection (hereafter Carl Taylor Collection), Box 21, Folder 'Narangwal General'.

[66] Johns Hopkins University Department of International Health, 'Health and Family Planning; Annual Report, 1969/1970' (1970): 17, USAID Document Experience Clearinghouse (hereafter USAID DEC), http://pdf. usaid.gov/pdf_docs/PNAAE757.pdf (accessed 15 November 2013).

[67] Johns Hopkins University Department of International Health, 'Health and Family Planning; Annual Report, 1968/69' (1969), USAID DEC, http:// pdf.usaid.gov/pdf_docs/PNAAE759.pdf (accessed 15 November 2013).

In summary, these early reports from the Narangwal population project identified 'systematic family planning' as a means to extend family planning to as many 'eligible' couples as possible. The Narangwal study staff developed their field procedures to achieve this through an approach which combined health care with family planning services. Yet, in the 1983 monograph of the Narangwal population project, Taylor and his colleagues claimed that the high percentage of 'new recruits' to family planning achieved within the Narangwal study was 'the most evident shift toward equity' in their findings.[68] In his 'Surveillance for Equity' article, Taylor claimed that such a 'systematic' approach to family planning could enable public health practitioners to control population growth while also focusing on 'equity'. He wrote:

> Deaths of children are mostly in poor families with the least educated mothers and the most babies… when a child's life is saved parents are helped to see that they do not need to have many children and that family planning can help increase the chances for child survival. Surveillance for equity can focus synergistic health and family planning services on needy families with the most children.[69]

Therefore, in the case of family planning, 'Surveillance for Equity' layered the language of social justice on top of a structure of surveillance which served principally to control population growth by targeting the bodies of the poor.

Risk

Early reports from the Narangwal study also explain how systematic monitoring enabled health workers to make more efficient interventions by identifying and targeting 'risk'. These reports explain how regular monitoring of the Narangwal study population was crucial to the principle of chiefly delivering health services though auxiliary health care workers. Or, as one report put it, 'maximum delegation of responsibility to those with minimum training consistent with good care'.[70] As part of this strategy, auxiliary health care workers aimed to identify 'high risk' patients for referral to a physician. For example,

---

[68] Taylor et al., *Integrated Family Planning and Health Care*, 49–50.

[69] Taylor, *International Journal of Epidemiology* 21, no. 6 (1992): 1045.

[70] Narangwal RHRC, *Population Project Annual Report, 1971–1972*, 85.

through regular surveillance, health care workers endeavoured to identify 'high risk' women who might need specialist medical care during childbirth. This approach allowed the remaining—less 'risky'—cases to be dealt with by auxiliaries or Traditional Birth Attendants, while still allowing for the improvement of overall perinatal and neonatal mortality.[71] The annual report of the Narangwal population project for 1971–2 explained that, 'If patients are contacted in their homes rather than awaiting their attendance at a clinic later in the disease when morbid processes are less amenable to the treatments of the auxiliary, medical care can successfully be delegated to non-medical members of the health team'.[72] Thus, the strategy of monitoring and targeting 'risk' helped to maximize health outputs at minimal cost by enabling the delegation of the majority of care to auxiliaries.

Monitoring 'risk' also allowed for the minimization of health care inputs in several other ways. For example, in 1973, Taylor explained the use of surveillance to monitor 'the high risk child' in the following terms:

> The fundamental principle underlying all our work is not to dissipate effort by providing mass care but to concentrate on developing surveillance techniques which can be used to focus attention on those in greatest need. Our nutrition program specifically concentrates on regular weighing to identify nutritional faltering and then provide intensive care when needed.[73]

The system of surveillance provided by Narangwal study's surveillance tactics thus allowed the study staff to reduce the cost of nutritional supplements by directing them only to those at highest risk. By contrast, in 1983 the authors of the Narangwal study described this same strategy as 'Mechanisms ... to ensure equity in distribution' of nutrition and health care.[74]

Surveillance was therefore a prominent feature of earlier reports from the Narangwal study. However, in those reports, Narangwal study staff presented surveillance as a mechanism to ensure maximum family

[71] Narangwal RHRC, *Annual Report of Population Research, 1972–73*, 23.

[72] Narangwal RHRC, *Population Project Annual Report to the ICMR 1971–1972*, 100.

[73] Narangwal RHRC, *Annual Report of Population Research 1972–73*, 21.

[74] Kielmann et al., *Integrated Nutrition and Health Care*, vol. 1, 1983, 64.

planning coverage, or to target 'risk'. Above all, the language of these reports is one of efficiency, and not of equity. From the late-1970s, Taylor and his colleagues represented this surveillance-based methodology as a path to social justice in health.

THE POLITICS OF HEALTH AND SOCIAL JUSTICE: FROM BUCHAREST TO ALMA ATA

How do we explain this shift in terminology? I suggest that Taylor and his colleagues' concern for 'equity' was a response to the prevailing political climate. Taylor began to write frequently about social justice and equity in the mid-1970s. He did so specifically in response to the questions surrounding international inequality, and the call for the redistribution of resources, represented by the New International Economic Order (NIEO). In 1974, a special session of the General Assembly of the United Nations adopted a declaration which called for the establishment of a NIEO 'based on equity, sovereign equality, interdependence, common interest and cooperation among all States … which shall correct inequalities and redress existing injustices'.[75] The General Assembly also adopted a Programme of Action which sought to implement the principles of the NIEO through a series of economic measures—incorporating international production, consumption and trade.[76]

The political climate surrounding the formation of the NIEO had a palpable impact on international health and population debates. One notable example of this effect was the rhetoric of the 1974 World Population Conference in Bucharest. During the Bucharest conference,

[75] United Nations General Assembly, 'Resolution adopted by the General Assembly, 3201 (S-IV). Declaration on the Establishment of a New International Economic Order', www.un-documents.net/s6r3201.htm (accessed 15 November 2013), 1; Director-General for Development and International Economic Co-operation, *Towards the New International Economic Order* (New York: United Nations, 1982), 3.

[76] United Nations General Assembly, 'Resolution adopted by the General Assembly, 3202 (S-VI). Programme of Action on the Establishment of a New International Economic Order', www.un-documents.net/s6r3202.htm (accessed 15 November 2013), 1–16; Director-General for Development and International Economic Co-operation, *Towards the New International Economic Order*, 4.

delegates from the so-called third world nations rejected the idea of narrow population control measures in favour of a broader approach based on the redistribution of resources and economic development.[77] For instance, Karan Singh, Minister of Health and Family Planning for India, declared that 'development is the best contraceptive': a move which—like the ostensible drive for health 'equity'—cannot be seen as anything but rhetorical in light of the aggressive population control programme which Singh oversaw during the Emergency of 1975–7.[78]

In his international health writings, Taylor noted this increasing concern for social justice and human rights, and referred specifically to the NIEO and Bucharest conference.[79] Taylor wrote that the NIEO was 'attempting to articulate a new approach to development focusing on the basic human needs of the world's poorest billion'.[80] He argued that international assistance programmes should 'help translate these concepts of social justice into international standards and norms'.[81] Through his involvement with 'equity', therefore, Taylor simply reinterpreted his pre-existing public health work in the light of prominent political concerns.

Others also began to talk about health 'equity' during this period. Indeed, during the 1970s, the concept of 'equity' in health began to take off internationally. Figure 3.1—which shows the number of texts which employed the words 'health' and 'equity' between 1970 and 2013 in the PubMed database of biomedical literature—illustrates this rise.[82] From occasional references in the early 1970s, use of the term increased rapidly towards the close of the decade.

[77] Jason L. Finkle and Barbara B. Crane, 'The Politics of Bucharest: Population, Development, and the New International Economic Order', *Population and Development Review* 1, no. 1 (1975): 87–114.

[78] Finkle and Crane, 'Politics of Bucharest', 106; Rebecca Williams, 'Storming the Citadels of Poverty: Family Planning under the Emergency in India, 1975–77', *Journal of Asian Studies* 70, no. 2 (2014).

[79] Carl E. Taylor, 'Changing Patterns in International Relationships: Motivation and Relationships', *American Journal of Public Health* 69, no. 8 (1979): 805; Taylor, 'Economic Triage of the Poor and Population Control', *American Journal of Public Health* 67, no. 7 (1977): 660.

[80] Taylor, 'Changing Patterns in International Relationships', 805.

[81] Taylor, 'Changing Patterns in International Relationships', 805.

[82] PubMed—http://www.ncbi.nlm.nih.gov/pubmed—is the database of biomedical literature of the United States National Library of Medicine, and the world's largest database of medical citations and abstracts.

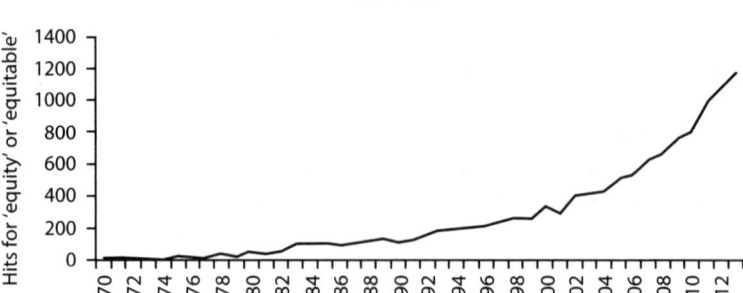

FIGURE 3.1   Number of texts per year containing the words 'equity' or 'equitable' in the United States National Library of Medicine's PubMed database of biomedical literature

Source: Data from http://www.ncbi.nlm.nih.gov/pubmed (accessed 21 November 2013).

A more detailed look at medical journals revealed a distinct shift in the use of the word 'equity' during the 1970s. Towards the start of that decade, most articles which used the word 'equity' used the term to connote *wealth*. That is, 'equity' in the sense of economic assets: 'equity capital' in private sector hospitals; 'equity' investments in medical group practices; 'equity' in savings deposits in the context of discussions on health insurance.[83] By the mid-1970s, authors more frequently used the term 'equity' in the sense of social justice or equality.

The declaration of 'health for all' which the International Conference on Primary Health Care made at Alma Ata in 1978 was also part of this historical move towards health 'equity'.[84] At the Alma Ata conference,

[83]   Charlotte Muller, 'Health at What Price? Some Notes for Comprehensive Health Planners', *American Journal of Public Health* 59, no. 4 (1969): 655; E. Sidney Willis, 'Medical Care for Industrial Workers—Manager's Viewpoint', *American Journal of Public Health* 47, no. 4 (1957): 461; George Baehr, 'The Family Physician as the Central Figure in Prepaid Group Practice', *American Journal of Public Health* 43, no. 2 (1953): 135.

[84]   Sudhir Anand and Fabienne Peter, 'Introduction', in *Public Health, Ethics, and Equity*, 2.

international health experts attempted to address inequalities in access to health care through a shift to a model of Primary Health Care.[85] The background documents to the Alma Ata conference—which Taylor helped to author—also explicitly cited the NIEO as part of the rationale for tackling these inequalities. Paragraph three of the Alma Ata declaration itself stated that 'Economic and social development, based on a New International Economic Order, is of basic importance to the fullest attainment of health for all and to the reduction of the gap between the health status of the developing and developed countries.'[86] Therefore, the concept of 'health for all' was underpinned by the broader idea of international social justice embodied by the NIEO. Echoing the calls for equitable redistribution, the Alma Ata declaration called for a 'transfer of a greater share of health resources to the underserved majority of the population'.[87] The declaration also demanded a 'more equitable distribution of international health resources to enable the developing countries ... to apply primary health care'.[88]

The Depoliticization of Poverty

By contrast, Taylor's concept of 'surveillance for equity' bypassed these commitments to redistribution entirely. In fact, 'surveillance for equity' was designed to produce optimal health 'outputs' at low cost, and without additional input of resources or economic redistribution. Indeed, Taylor argued that in countries with 'extreme financial constraints', providing additional health services would contribute less than 'reaching families with the greatest health problems'. Taylor argued that simply adding resources would not ensure that services reached the

[85] Marcos Cueto, 'The Origins of Primary Health Care and Selective Primary Health Care', *American Journal of Public Health* 94, no. 11 (November, 2004): 1864–74.

[86] World Health Organisation (WHO) and the United Nations Children's Fund (UNICEF), *Primary Health Care: Report of the International Conference on Primary Health Care, Alma-Ata, USSR, 6–12 September 1978* (Geneva: WHO, 1978), 78; WHO, *Declaration of Alma-Ata: International Conference on Primary Health Care*, Alma-Ata, USSR, 6–12 September 1978, http://www.who.int/hpr/NPH/docs/declaration_almaata.pdf (accessed 17 March 2012).

[87] WHO and UNICEF, *Primary Health Care*, 12–13.

[88] WHO and UNICEF, *Primary Health Care*, 12–13.

poor, who would continue to suffer disproportionately from their more 'easily reached and convinced' neighbours. In the monographs of the Narangwal study, Taylor and his colleagues suggested that the poor, unlike more prosperous groups, 'do not spontaneously seek health care', even where free services were available, 'because of a long tradition of psychological, geographical, and social barriers'.[89] Economic barriers were notably absent from this explanation. Thus, the authors of the Narangwal study concluded that, 'Outreach is needed to overcome the long-standing reluctance of poor people to open themselves to the possibility of being rebuffed if they ask for help'.[90] Surveillance could help health workers to identify and respond to problems, and allow for more 'equitable coverage'.[91]

Crucially, 'surveillance for equity' was to provide aid agencies and policy makers a means to achieve health outputs *without* corresponding social or economic equality. Indeed, in a discussion paper he wrote for the United Nations Children's Fund, Taylor stated that, 'We have to go beyond idealistic pronouncements about equalizing access and intangible goals such as social justice.' Instead, they should focus on specific 'equity' indicators to target those in greatest need for maximum impact at minimum cost.[92]

Taylor drew upon two examples in support of his health equity-without-economic equality thesis: Kerala and Sri Lanka. He observed that in both those places, governments had provided nearly universal coverage of health care, family planning, education, and basic nutrition, without any increase in economic output. Describing this process in Kerala, Taylor stated enthusiastically:

> Per capita income is close to the lowest among Indian states. The fertility decline indicates what can be done by aggressively providing health services and education, and more recently family planning, even without the economic growth that has characterised the richer states of India where birth rates are also declining.[93]

---

[89] Taylor et al., *Integrated Family Planning and Health Care*, 50.

[90] Taylor et al., *Integrated Family Planning and Health Care*, 50.

[91] Taylor et al., *Integrated Family Planning and Health Care*, 2.

[92] Carl Taylor, *Issues for Discussion in UNICEF*, Carl Taylor Collection, Box 10, Folder Correspondence 1993.

[93] Taylor, 'Economic Triage of the Poor', 662.

Taylor's message was that policy makers need not alleviate poverty and injustice in order to achieve health 'equity'. Rather, they could achieve ostensibly 'equitable' health outcomes even with prevailing inequalities.

In this same vein, Taylor and his Narangwal study colleagues noted that improvements in the health and nutritional status of children typically occurred 'spontaneously as a result of general socioeconomic development'. However, they claimed that the Narangwal nutrition study had demonstrated that 'it is not necessary now to wait for general development, since measures are available that contribute directly to improving the quality of children's lives'.[94] Therefore, through a series of technical interventions, health experts could address poor health without addressing poverty.

Indeed, the Narangwal study was geared towards providing practical solutions to the physical manifestations of poverty. The design of the Narangwal study incorporated strategies such as the recruitment of individuals who had previously worked on practical development programmes serving poor communities. Reports from the Narangwal study stressed that such strategies helped to keep services 'simple, practical and inexpensive' by reducing 'the tendency to concentrate on complicated and esoteric ideas which seem academically interesting but are beyond the economic constraints of developing countries'.[95] In other words, the Narangwal study deliberately circumvented the problem of poverty in order to focus on specific, technical interventions which were within the immediate remit of public health experts.

Entering Wedge

Yet, Taylor did view these short-term technical interventions as part of a longer-term strategy of economic transformation. Taylor claimed that the integrated package of health, family planning, and nutrition which he proposed to deliver via the 'surveillance for equity' model could not only deliver improved health outcomes, but also act as 'an

[94] Taylor et al., *Integrated Family Planning and Health Care*, 63.

[95] Narangwal RHRC, 'The Narangwal Population Study: Integrated Health and Family Planning Services' (1975), I.A.33, Carl Taylor Collection, Box 1.

entering wedge in the process of socioeconomic development'.[96] It is striking that Taylor chose this particular phrase—'entering wedge'—to talk about health care provision. It is striking because this is the same phrase which the Rockefeller Foundation used to describe the role of its public health work in paving the way for industrial capitalism during the early twentieth century. Wickliffe Rose, the first director of the Rockefeller Foundation's International Health Board (IHB), described the IHB's hookworm campaigns as an 'entering wedge' to enliven public interest and participation in public health.[97] Hookworm—otherwise known as 'the germ of laziness' because of its association with low productivity—was the Rockefeller Foundation's first public health project. The Rockefeller Sanitary Commission instigated its first hookworm campaign in 1909 in order to facilitate productive industrial capitalism in the southern states of the US.[98] Internationally, another Rockefeller Foundation officer described their public health work abroad as 'an entering wedge for permanent civilizing influences' such as industrial schools.[99]

Just as the Rockefeller Foundation's earlier public health work aimed to pave the way for industrial capitalism, Taylor saw 'surveillance for equity' as a means to raise the productivity of the poorest peoples in the 'underdeveloped' nations and pave the way for economic 'development.' He wrote:

> It seems evident that in order to work hard people need hope. If life holds less threat of death, illness, chronic weakness from lack of nutrients, and days lost from preventable mass diseases the daily efforts of the poor will be more productive. If children grow up without having been deprived of maternal attention and food because the family has more

[96] Taylor, 'Economic Triage of the Poor', 663; Carl Taylor, 'Foreword' to Mabelle Arole and Rajanikant Arole, *Jamkhed: A Comprehensive Rural Health Project* (Ahmednagar: Comprehensive Rural Health Project, 1994), 7.

[97] Anne-Emanuelle Birn and Armando Solórzano, 'Public Health Policy Paradoxes: Science and Politics in the Rockefeller Foundation's Hookworm Campaign in Mexico in the 1920s', *Social Science and Medicine* 49 (1999): 1198.

[98] Anne-Emannuelle Birn, 'Public Health or Public Menace? The Rockefeller Foundation and Public Health in Mexico, 1920–1950', *Voluntas* 7, no. 1 (1996): 38.

[99] Cited in E. Richard Brown, *Rockefeller Medicine Men: Medicine and Capitalism in America* (University of California Press, 1979), 124.

children than can be cared for, they will have measurably better intellectual and physical development and a greater chance of growing into adults who contribute to progress.[100]

Similarly, in the monograph of the Narangwal nutrition study, Taylor and his colleagues argued that the 'optimum development of children', in particular, 'represents any society's greatest resource for the future'.[101] In other words, Taylor viewed international health projects not only as a way to improve the health of the poor, but to transform them into a productive workforce.

★ ★ ★

As I noted earlier in this chapter, the word 'equity' carries more than one meaning. It can mean justice, impartiality, and fairness. However, it can also mean an economic investment. Thus, I argue that 'surveillance for equity' only makes sense if we think of 'equity' not in terms of fairness of justice, but in terms of wealth. Within the 'surveillance for equity' model, the health of the people is the wealth of the nation.

Though primarily concerned with the health of the poor, 'surveillance for equity' functioned to sideline the political and structural causes of poverty in favour of a narrow focus on health outcomes. In viewing the diseases of poverty as a series of technical problems requiring technical solutions, Taylor's approach to 'equity' in health depoliticized the problem of poverty. While formulated in response to international demands for economic redistribution, 'surveillance for equity' was not a new approach to health care. Rather, it was a way to rationalize health services in order to achieve optimal health outputs at minimal cost. In place of socioeconomic justice, the poor were provided with surveillance.

Since Taylor formulated his approach to 'equity' in health, international health projects have continued to implement systems of continuous monitoring in an attempt to deliver equitable health outcomes.[102] Yet, as Fabienne Peter and Timothy Evans argue, if health

[100] Taylor, 'Economic Triage of the Poor', 662.

[101] Kielmann et al., *Integrated Nutrition and Health Care*, vol. 1, 1983, 3.

[102] Berggren et al., 'Longitudinal Community Health Research, 157–8; P. Freeman, H.B. Perry, S.K. Gupta, and B. Rassekh, 'Accelerating Progress

inequalities are rooted in broader social processes, then the remedies to those inequalities must also be sought in 'social policies at large'.[103] In other words, health 'equity' cannot function as an apolitical concept, with a narrow focus on health outcomes and specific, technocratic interventions. Rather, the search for equity in health must be embedded in a broader pursuit of equality.[104]

---

in Achieving the Millenium Development Goal for Children through Community-based Approaches', *Global Public Health* 7, no. 4 (April, 2012): 411; Henry B. Perry, Leslie W. King-Schultz, Asma S. Aftab, and John H. Bryant, 'Health Equity Issues at the Local Level: Socio-geography, Access, and Health Outcomes in the Service Area of the Hôpital Albert Schweitzer-Haiti', *International Journal for Equity in Health* 6, no. 7 (2007), http://www.equityhealthj.com/content/6/1/7 (accessed 12 April 2012); Henry Perry et. al, 'Attaining Health for all through Community Partnerships: Principles of the Census-based, Impact-oriented (CBIO) Approach to Primary Health Care Developed in Bolivia, South America', *Social Science and Medicine* 48, no. 8 (1999): 1053–67; Khatidia Huesin et al., 'Developing a Primary Health Care Management Information System that Supports the Pursuit of Equity, Effectiveness and Affordability', *Social Science and Medicine* 36, no. 5 (1993): 585–96.

[103] Fabienne Peter and Timothy Evans, 'Ethical Dimensions of Health Equity', in Timothy Evans et al., *Challenging Inequities in Health*, 31.

[104] Evans et al., *Challenging Inequities in Health*, 26; Fabienne Peter, 'Health Equity and Social Justice', *Journal of Applied Philosophy* 18, no. 2 (2001): 159–70.

Part 2

India's Hospitals: For Whom?

4 Globalization and the Health of a Megacity

The Case of Mumbai

Ramila Bisht and Altaf Virani

The state (public partner), citizens (beneficiaries), and private enterprises (concessionaires) are forging a new relationship in a rapidly globalizing environment, called public–private partnership or PPP. With dwindling public resources and poor efficacy in delivering public good, even social policy—a quintessential duty and responsibility of the state—is riding this new vehicle. However, a PPP in health care in Mumbai has gone all wrong for the 'beneficiaries', especially the poor and the marginalized. The chapter reviews the health care landscape of Mumbai over the last few eventful decades and examines the PPP that converts a dysfunctional municipal maternity home into a multi-specialty peripheral hospital, ostensibly to benefit the slum-dwellers in the catchment area. It further attempts to infer the implications of PPPs on access to health care services and draws lessons for public policy from this experience.

With global economic and market forces increasingly driving social policies, erstwhile relationships between the democratic welfare state and its citizens are getting radically redefined with the private sector playing a significantly greater role. For countries low on human development, opening up of social sectors like health and education raises serious concerns about the efficacy of such measures. These policy shifts have raised moral dilemmas and ideological debates the world over.

In India, international norms of health policy—with greater emphasis on the private sector—were mediated through the World Bank package for health reforms initiated during the 1990s. International debates on health strategies for developing countries pushed policy towards privatization of public provision for health care through PPPs. A central concern with PPPs as a prominent route to organizing and delivering public health services is its suitability and effectiveness in honouring public health commitments of the state. What do such partnerships mean for equity and access to good health care services for the poor in a context where trends in public health are not very encouraging? We will examine the withering social commitment of the state and the resulting inequities through an analysis of changing modes of health service delivery, particularly the emergence of PPPs.

This chapter attempts to assess the social impact of the changing health care delivery system in the city of Mumbai in Maharashtra. It is divided into three sections. The first provides an overview of Mumbai's health care system, taking stock of the public and private health infrastructure and the inequities it breeds. The second section studies the changes in the city's health care infrastructure over the last two decades with a focus on health care for the poor. It also reviews how new policy initiatives push towards privatization of the public health system and how these changes are formulated and operationalized. In the third section, the first public health care PPP of its kind in Mumbai, in which a Municipal Corporation of Greater Mumbai (MCGM) maternity home underwent a transformation, is discussed as an illustration of changing relationships of state and private sector. In conclusion, the broader implications of the findings of this case study for the poor are analysed in the context of engendering greater exclusions through new forms of marginalization.

WHY MUMBAI?

Mumbai, including the island city and its suburban districts, currently has a population of 12.5 million with a density of 20,482 persons per square kilometer.[1] India's financial capital with its economic advance, efficient governance, and multicultural character, is seen as an international city that has experienced exponential economic growth. Its large public health system is one of the best in the country. However, despite a relatively vast supply of public hospitals, the basic health needs of large swathes of the population remain unmet. With these characteristics, Mumbai is a good case for studying efforts towards market integration of the health system.

FAR FROM EQUITABLE ACCESS

With its large diversified network of public and private health institutions that represent a typical concentration of urban health services in India, Mumbai has a disproportionately higher concentration of health care infrastructure per capita compared to the rest of Maharashtra. In all, Mumbai has five teaching hospitals, 1,492 general hospitals, 176 primary Health Centres and health posts, 2,067 dispensaries, an aggregate bed strength of 43,902 and a somewhat respectable bed-population ratio of 282 beds per 100,000 persons—the highest in the state, as detailed in Table 4.1.

Till the 1950s, almost three-quarters of Mumbai's 50-odd hospitals were public. The rest were mostly managed by not-for-profit trusts. By the 1980s, the private sector had begun to establish itself as a significant health care provider—especially to the middle and upper income groups. This included the voluntary sector, non-governmental organizations, small clinics, nursing homes, 'charitable' trust hospitals, and large corporate groups. On the other hand, the government was focused on setting up or rejuvenating health posts, dispensaries, post-partum centres, and maternity homes through the World Bank-funded India Population Project.[2] As the government's attention was shifting away from curative care, growth in secondary and tertiary public

[1] Census of India, 2011 (New Delhi: Office of the Registrar General and Census Commissioner of India, 2011).

[2] Loraine Kennedy, Ravi Duggal, and Stephanie Tawa Lama-Rewal, 'Assessing Urban Governance through the Prism of Healthcare Services in Delhi, Hyderabad and Mumbai', in *Governing India's Metropolises*, ed. Joël Ruet and Stéphanie Tawa Lama-Rewal (New Delhi: Routledge, 2009), 174.

TABLE 4.1   Public and private facilities in Mumbai's health system

Type of Facility	Public Facilities	Private Facilities	Total
Teaching Hospitals	4	1	5
General Hospitals	76	1,416	1,492
Health Posts	176	–	176
Dispensaries	235	1,832	2,067
Hospital Beds	20,700 (29% of beds in Maharashtra)	23,202 (37% of beds in Maharashtra)	43,902

Source: Adapted from Ravi Duggal, T.R. Dilip, and Prashant Raymus, *Health and Healthcare in Maharashtra: A Status Report* (Mumbai: CEHAT, 2005). The data is multitemporal and compiled from many sources.

health infrastructure diminished, making way for the private sector to provide curative services. By the 1990s, public hospitals were grossly underfinanced and had begun to dilapidate. Although the private sector occupies an eminent position in the delivery of health care services in the city today, the public health system, albeit weakened, continues to be the mainstay for a large section of the population. A majority of the public health services are provided by the Municipal Corporation of Greater Mumbai (MCGM) and the state government.[3] The public health system consists of a multi-tiered health network (see Table 4.2) that treats millions of patients each year drawn from within Mumbai city as well as outside.

Over the years Mumbai's peculiar geography and factors influencing its economic transformation—the rise of the tertiary sector, the fragmentation of production, and the rise of the informal sector—have defined its spatial growth.[4] However, the city's trajectory of demographic, economic, and physical growth and expansion has not been complemented by appropriate and aligned expansion of the structure of health care provision and its distribution.

Continuous in-migration has resulted in a greater proportion of the population settling towards the north along the suburban rail net-

---

[3] After Bombay's renaming, the Bombay Municipal Corporation (BMC) was renamed Municipal Corporation of Greater Mumbai (MCGM).

[4] Arun Kumar Acharya and Praveen Nangia, 'Population Growth and Changing Land-Use Pattern in Mumbai Metropolitan Region of India', *Caminhos de Geografia* 11, no. 11 (2004): 168–85.

TABLE **4.2**   Public health infrastructure in Mumbai

Municipal Corporation of Greater Mumbai	Government of Maharashtra
Primary Care	*Secondary Care*
Health Posts (168)	General Hospitals (3)
Dispensaries (162)	*Tertiary Care*
Post-partum Centres (24)	Teaching Hospitals (2)
Secondary Care	
Peripheral Hospitals (16)	
Specialized Hospitals (5)	
Maternity Hospitals (27)	
Tertiary Care	
Teaching Hospitals (4)	

Source: Adapted from Municipal Corporation of Greater Mumbai (MCGM), *Mumbai City Development Plan 2006* (Mumbai: MCGM, 2006); and A.K. Jain, *Public Health Infrastructure in Mumbai* (Mumbai: Mumbai Transformation Support Unit, All India Institute of Local Self Government, 2006).

work and peripheral areas of the Mumbai Metropolitan Region (see Figure 4.1).[5] As per the Census of 2011, merely a quarter of the population lives in the island city or South Mumbai. However, public health care services are concentrated in South Mumbai, which has 63 per cent of public hospital beds in the city. On the other hand, Mumbai's suburbs have only 37 per cent of the total public hospital beds (see Table 4.3). Mired in political, economic, and demographic factors, the expansion and distribution of public health services has been lopsided.

Inequity also plagues the commitment of public health services to the poor. A majority of the 5.8 million slum-dwellers of Mumbai reside in suburbs, compounding the negative impact of public under-provisioning. Slum-dwellers, accounting for 54 per cent of the total population of western suburbs and an overwhelming 74 per cent of eastern suburbs, have access to only 20 per cent and 17 per cent of municipal hospital beds respectively. In comparison, a third of the South Mumbai population that lives in slums can access 63 per cent of the municipal hospital beds in South Mumbai as shown in Table 4.4.

[5] Mumbai Metropolitan Region (MMR) covers 4,355 square kilometres and includes the island city, western and eastern suburbs (together forming Greater Mumbai), Navi Mumbai, and parts of Thane and Raigad spanning seven major civic corporations.

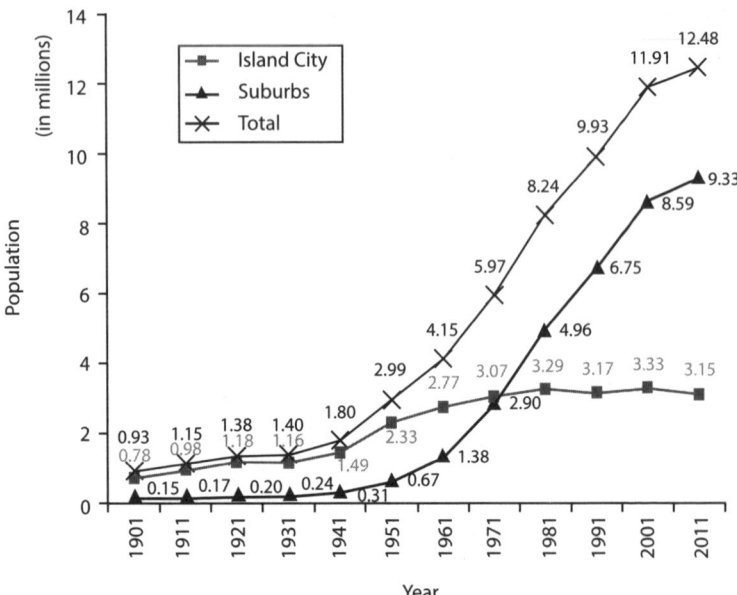

FIGURE 4.1   Population trends in Greater Mumbai
Source: Compiled from Census of India, 'General Population Tables for Greater Mumbai Municipal Corporation' (New Delhi: Office of the Registrar General and Census Commissioner of India, 1901–2011).

TABLE 4.3   Region-wise distribution of hospital beds in Mumbai

Region	Population	Municipal Beds	Others Beds	Total
South Mumbai	3,700,098	6,386	13,577	19,963
	(28%)	(63%)	(50%)	(53.3%)
Western Suburbs	5,689,012	2,059	8,972	11,031
	(43%)	(20%)	(33%)	(29.5%)
Eastern Suburbs	3,888,610	1,702	4,723	6,425
	(29%)	(17%)	(17%)	(17.2%)
Total	13,277,720	10,147	27,272	37,419
	(100%)	(100%)	(100%)	(100%)

Source: Public Health Department, MCGM, *Yearly Report, 2006*; MCGM, *Mumbai Human Development Report 2009*, 122.

TABLE 4.4   Region-wise distribution of municipal hospital beds vis-à-vis
slum populations in Mumbai

Region	% of Slum Population in Each Suburb	Municipal Beds	% of Beds in Municipal Hospitals
South Mumbai	32.73	6,386	63
Western Suburbs	54.02	2,059	20
Eastern Suburbs	74.41	1,702	17
Total		10,147	100

Source: MCGM, *Yearly Report, 2006*; Census of India, 'General Population
Tables for Greater Mumbai Municipal Corporation' (New Delhi: Office of the
Registrar General and Census Commissioner of India, 2001); MCGM, *Mumbai
Human Development Report 2009*, 69, 122.

Even this limited capacity of the public health system is highly
skewed away from primary and secondary health care needs, thus creating an imbalance in services. Almost 42 per cent of municipal hospital
beds are concentrated in the city's apex institutes in South Mumbai.
Peripheral hospitals offering the much needed general services make
up less than 36 per cent, while maternity homes constitute a meagre 5
per cent of the total bed strength in civic hospitals.[6]

THE CURRENT SCENARIO

Private hospitals in Mumbai account for 53 per cent of the city's
aggregate hospital bed-strength.[7] With the corporatization of health
care services, hospitals are regularly being established with share capital
and borrowings. A primarily curative private sector health service is
now catering to every social stratum with its range of offerings. Driven
by a quest for profits, only services of inferior quality are accessible to
the poor, who end up paying for sub-standard private medical care.[8]

[6] MCGM data cited in Jain, *Public Health Infrastructure in Mumbai*.

[7] Duggal, Dilip, and Raymus, *Health and Healthcare in Maharashtra*, 16.

[8] Padma Bhate-Deosthali, Ritu Khatri, and Suchitra Wagle, 'Poor Standards
of Care in Small, Private Hospitals in Maharashtra, India: Implications for
Public–private Partnerships for Maternity Care', *Reproductive Health Matters*
19, no. 37 (2011): 32–41.

The public health care system is seriously overstretched by corruption, patient–overload, and shortage of health workers, medical services, medicines, and financial resources, resulting in 32 per cent ailments going untreated.[9] Instead of the seven million it is provisioned to serve, it currently serves more than 13 million. With ill-located facilities that are unable to optimize the triad of costs, waiting-time, and treatment for the poor, the country's richest municipality caters to barely 20 per cent of its population despite spending 25 per cent of its budget on health.[10] In spite of these inadequacies, the poor mostly seem to prefer public hospitals and avail private health care only in the absence of public health services.[11] Against this backdrop of unmet demand and rising despondency of the poor, the state government of Maharashtra and MCGM decided to bring in PPPs in health care as a potential solution.

PUBLIC–PRIVATE PARTNERSHIPS IN MUMBAI'S HEALTH CARE SYSTEM

Compelled by a severe budget deficit in the late 1990s, MCGM was forced to charge user fees in its public hospitals. In doing so, MCGM transferred its financial burden to the poor.[12] It also took World Bank

---

[9] Sunil Nandraj, Neha Madhiwalla, Roopashri Sinha, and Amar Jesani, *Women and Health Care in Mumbai* (Mumbai: CEHAT, 2001), cited in T.R. Dilip and Ravi Duggal, *Demand for Public Health Services in Mumbai* (Mumbai: CEHAT, 2003); T.R. Dilip and Ravi Duggal, 'Unmet Needs for Public-Health Care Services in Mumbai, India', *Asia-Pacific Population Journal* 19, no. 2 (2004): 27–40.

[10] Dilip and Duggal, 'Unmet Needs for Public-Health Care Services in Mumbai, India', 29.

[11] Kennedy, Duggal, and Lama-Rewal, 'Assessing Urban Governance', 161–82; Duggal, Dilip, and Raymus, *Health and Healthcare in Maharashtra*; Dilip and Duggal, 'Unmet Needs', 27–40; Nandraj et al., *Women and Health Care*.

[12] The policy implemented in 1998, charged patients for papers, tests, medicines, surgeries, and so on. See Dilip and Duggal, *Demand for Public Health Services in Mumbai*; Dilip and Duggal, 'Unmet Needs'; Duggal, Dilip, and Raymus, *Health and Healthcare in Maharashtra*; Kennedy, Duggal, and Lama-Rewal, 'Assessing Urban Governance'.

advice and instituted the PPP model in health care with the stated aim of 'effective and efficient delivery of quality social services to the people'.[13] Initially limited to hiring in or contracting out services, MCGM eventually took the PPP route to try and revitalize dying municipal hospitals.[14] Guidelines for tendering, scrutinizing, and contracting private partners to run health facilities on 'care-taker' and 'no-profit-no-loss' basis for a period between 10 and 30 years provided that the facilities be kept free for the poor while charging the paying patients 'reasonably'.[15]

While pushing for PPPs, MCGM has offered numerous incentives to private players in the development, maintenance, running, and management of its dysfunctional hospitals. These moves have raised concerns and public ire, but have been met with either state apathy or tokenism. Despite its improved financial position, MCGM has neither established new hospitals nor upgraded existing ones.[16]

Public–private partnerships are of course not entirely new. They have existed in the past, although on a modest scale. Today however, they are not only ubiquitous, but also have a new meaning. The changing nature of PPPs today is placed within the larger context of changing infrastructural needs, which in turn is driven by new economic imperatives. Prime land, on which many public hospitals stand, is lucrative for the health entrepreneur–builder nexus that often enjoys tacit support from politicians and bureaucrats. While powerful business houses actively influence such decisions, international consultancy firms sanctify these through new models of delivering public health

---

[13] Kennedy, Duggal, and Lama-Rewal, 'Assessing Urban Governance', 174.

[14] This included hiring in or contracting out of both clinical and non-clinical services in public facilities to private agencies or individuals.

[15] *Indian Express*, 'Welfare Centre Run by Public Organizations', Mumbai, 19 June 2001; *Sakaal Times*, 'State to Come Out with PPP Policy', Mumbai, 28 January 2010.

[16] In 2002, MCGM with a surplus of INR 450 million allocated monies to transport and communication infrastructure, while primary health care budget was reduced. Capital expenditure on health dropped from 7.38 per cent (2007–8) to 5.07 per cent (2010–11) of total capital expenditure of the municipal corporation.

services.[17] The Chief Minister of Maharashtra's task force, in pursuing its grandiose vision to develop Mumbai into a world-class hub for high-end health care services, also joined the agenda.[18]

As a result, a new 'social policy' has evolved that seeks to make public services commercially viable, through virtual privatization of public assets via PPPs. Against this backdrop, the rest of this chapter will examine a specific PPP in Mumbai.

THE CASE OF SWAN MUNICIPAL GENERAL HOSPITAL

Methodology

This chapter is based on a study carried out between April and May 2010. The methodology included secondary research (newspaper articles, journal publications, reports, and so on) and primary data collection through observations, in-depth interviews, and group discussions. The respondents were purposively selected and included the management and staff of SWAN Municipal General Hospital (henceforth, SWAN MGH), the local corporator, MCGM officials, social activists, and beneficiary patients (and their families) who were treated at the hospital.[19] SWAN MGH officials agreed to a request for a one-time interview but

[17] Bombay First and McKinsey & Company, Inc., advised MCGM to bring reputed trusts and NGOs on PPP basis to manage hospitals without withdrawing funding, Bombay First and McKinsey & Company, Inc., *Vision Mumbai: Transforming Mumbai into a World-Class City—A Summary of Recommendations* (Mumbai: McKinsey & Company, Inc., 2003), 24. Bombay First and KPMG India Pvt. Ltd advised the government to get in the private sector for building new 'public' hospitals and refurbishing and managing existing ones for a fixed annuity with tax exemptions, subsidies, and additional FSI to expand services for paying clients, Bombay First and KPMG India Pvt. Ltd, *Healthcare: Mumbai Metropolitan Region* (Mumbai: KPMG, India Pvt. Ltd, 2009), 28–9.

[18] Government of Maharashtra, *Transforming Mumbai into a World-Class City: First Report of the Chief Minister's Task Force* (Mumbai: Government of Maharashtra, February 2004).

[19] To maintain anonymity, the identity of the hospital, its promoters, interviewed respondents, and their localities have been masked or not been disclosed.

were reluctant to either share concrete information (agreement details, utilization statistics, and beneficiary details) or allow patient-interviews in wards. So, beneficiaries from nearby slums were identified and interviewed through household visits. The partnership agreement and some statistics were obtained by filing an RTI application.[20]

Background

The 86-bedded hospital, well-utilized by women from eight nearby slums till the 1980s, was reduced to an effective bed strength of 20 by the mid-1990s owing to neglect by MCGM. In 1997–8, the state-owned power company offered to upgrade it to a 100-bedded peripheral hospital to provide health care for its employees.[21] In 1998, MCGM and the power company (henceforth PC) entered into an agreement that resulted in the reconstruction of a 544 square feet (basement + ground + six floors) peripheral hospital at a cost of INR 62.5 million using PC's welfare funds. However, MCGM soon declared its inability to keep its commitment of annual operational costs, estimated at INR 22.5 million. With PPPs in vogue, MCGM entered into a tripartite memorandum of understanding (MoU) with PC and a faith-based charitable trust (henceforth CT), in April 2002.

The Contours of the MoU[22]

The CT was entrusted with the maintenance and management of SWAN MGH—a 100-bedded hospital equipped with an out-patient department (OPD), an in-patient department (IPD), and an operation theatre; services in medicine, surgery, paediatrics, gynaecology and obstetrics, and basic diagnostic facilities in pathology and radiology—

[20] Right to Information (RTI) Act 2005 provides for replies to citizen requests for government information.

[21] According to MCGM and SWAN MGH officials, the strategic location of the hospital plot (then valued at INR 40 million, as per some estimates) was a critical criterion.

[22] Based on the tripartite agreement, 'Maintenance and Management Agreement for SWAN Municipal General Hospital' (name changed), dated 3 April 2002.

for an initial period of 30 years on 'caretaker' and 'no-profit-no-loss' basis.[23]

The CT's obligations:

1. To 'reserve' services and facilities for the poor and needy—40 per cent of all out-patients and 33 per cent of all in-patients—referred to as 'general' patients, at charges prescribed by MCGM.[24, 25]
2. To give quality care to all—ensure a high standard of medical care without discrimination and distinction between 'general' and paying patients.
3. To bear the full cost of running the hospital for which neither MCGM nor PC retain any financial or other obligations.

MCGM's obligations:

1. Allow CT unrestricted access—A 'quiet, exclusive, uninterrupted, unobtrusive and unhindered use' of the premises.
2. Allow CT to extend medical services to private patients to cross-subsidize 'general' patients.
3. Allow CT to develop additional infrastructure and services not covered under the reservation at its own cost including major construction using full Floor Space Index, subject to MCGM's approval.
4. Allow CT to raise funds through donations.
5. Fully exempt CT from paying general tax, education cess and street tax.
6. Provide CT subsidy on water, electricity and sewerage charges.
7. Make available all requisite 'No-Objection Certificates' to obtain licenses, permissions and approvals to successfully manage the hospital.

---

[23] It started as an 80-bedded facility, but grew to 100-bedded in a year's time as agreed in the MoU.

[24] Patients admitted in the general ward of SWAN MGH, which pertains to 33 per cent beds 'reserved' under the contract, are referred to as 'general patients' or 'reserved' category patients.

[25] Of the balance 67 per cent beds, 10 per cent are reserved for PC employees on preferential basis at rates approved by the advisory committee formed by three members each from MCGM, PC, and CT.

Advantage Concessionaire,[26] Disadvantage Poor

Prime land was allotted by MCGM to CT for a period of 30 years without any lease rent. The agreement allowed CT to build additional infrastructure and provide extra services for private patients, while availing tax exemptions and subsidy benefits from MCGM in return for providing nominal services to the poor. This kind of an arrangement virtually privatizes the public health facility, without adequate measures to make the private partner accountable to the government.

Given that the main policy intent behind existing PPP projects is better access to health care for citizens—particularly the poor, the question of how many beds are actually made available to service the poor merits close consideration. Within the MoU, key terms like 'poor' and 'needy' are left undefined. Although functionaries indicated that MCGM has directed the hospital to take in everybody who comes, in the absence of clear guidelines it appears that SWAN MGH uses arbitrary methods to decide who qualifies for these services. This often results in what some may consider 'poor' or 'needy' patients being forced to pay or turned away on the basis of income criteria defined by the hospital.[27]

As per the agreement, CT is required to provide all facilities and services available at SWAN MGH to 40 per cent out-patients and 33 per cent in-patients who are poor and needy, at the prevailing charges prescribed by MCGM for other government hospitals. However, out of the 33 per cent beds reserved for the poor and needy, 5 per cent is reserved for employees of MCGM and another five per cent for those of PC on preferential basis. From the remaining 67 per cent, another 10 per cent is reserved for PC on preferential basis. As a result, the actual number of beds available to the poor and needy is reduced to just 23 per cent (see Figure 4.2).

[26] Concessionaire in PPP is the private partner who designs, finances, builds, operates, maintains, transfers or does a combination/subset of these activities for a consideration or share in profit.

[27] SWAN MGH uses a medical social worker to screen patients for 'reserved' benefits on the basis of the BPL card, Tehsildar's certificate or ration card without clearly defined criteria. In practice 'reserved' benefits are denied to people with annual income above INR 25,000/- (threshold set by the Charity Commissioner in 2006), even if they cannot afford private services.

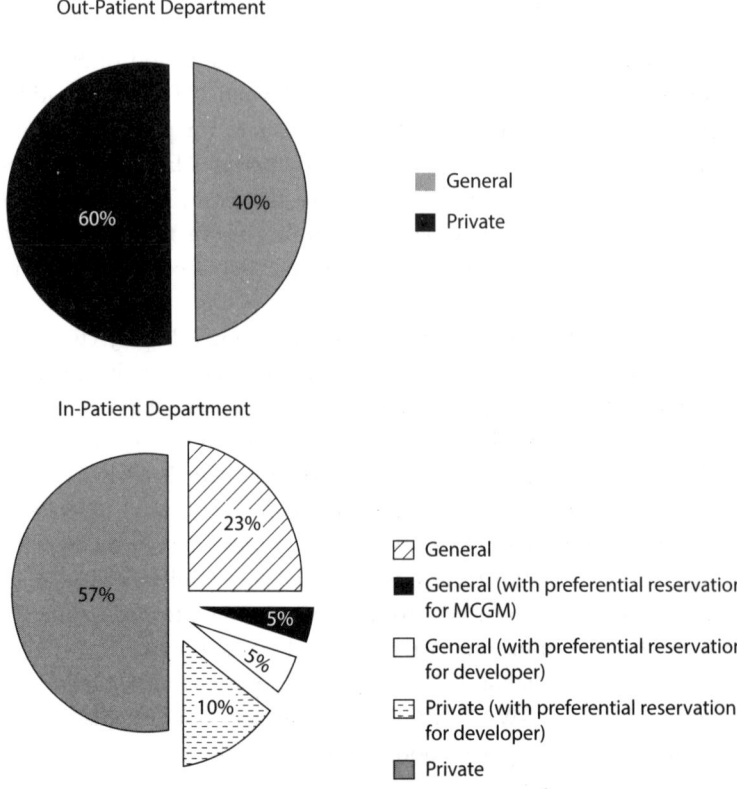

FIGURE 4.2   Distribution of services: who gets what and how much
Source: MCGM, PC, and CT, 'Maintenance and Management Agreement for SWAN Municipal General Hospital', Mumbai, 2002. This is a contractual agreement obtained from the Office of the Chief Medical Superintendent and HOD, Secondary Health Services, MCGM, under the Right to Information Act 2005. The names of two of the three parties to the contract (PC and CT) as well as the hospital (SWAN MGH) in the citation are pseudonyms for the sake of anonymity.

However, facility utilization data for private paying and 'reserved' categories of patients at SWAN MGH indicates non-adherence to even these allocations. In 2002–3 when the hospital first began its operations, general patients made up 78 per cent of the hospital's out-patient flow. Over the years, this proportion has gradually declined, hitting a low of 25 per cent in 2007–8. The share of 'reserved' category patients in the OPD stood at 34 per cent in 2009–10 as against the mandatory 40 per cent prescribed in the agreement. Similarly, the share of patients admitted in the IPD under the 'reserved' category has also dropped from over 53 per cent in 2002–3 to less than 32 per cent in 2009–10, falling short of the prescribed 33 per cent (see Figures 4.3a and 4.3b). Whereas the MoU clearly spells out that CT must provide subsidized services to 33 per cent of all in-patients (who are poor and needy) admitted to the hospital, both MCGM

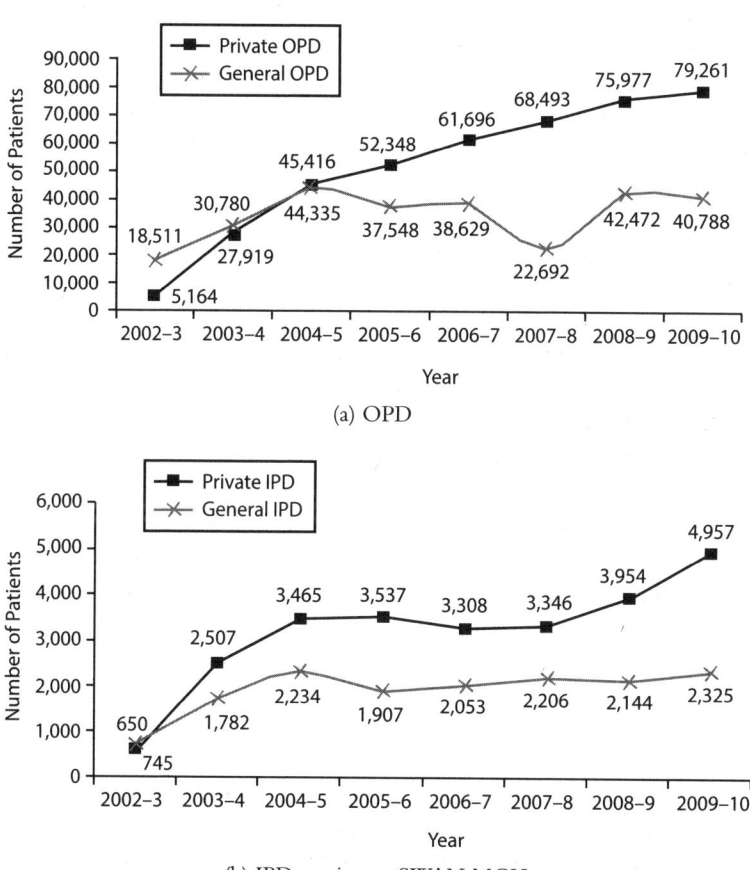

(a) OPD

(b) IPD services at SWAN MGH

FIGURE 4.3   Utilization statistics
Source: Data obtained from the Office of the Chief Medical Superintendent and HoD. Secondary Health Services, MCGM under Right to Information Act 2005.

and CT have interpreted this requirement as implying the allotment of 33 per cent of hospital beds for subsidized treatment of general category patients, leading to a further reduction in targets. Both parties have used this logic to argue that targets for IPD reservations have been consistently met over the years.

While CT seeks approvals for major modifications from MCGM, no serious control-mechanisms are embedded in the PPP. The CT-appointed hospital management board is practically the sole decision-maker for hospital operations, with strictly an advisory role for the committee with MCGM representation. Practically, MCGM exercises no executive control.

Soon after signing of the agreement, the management abbreviated 'Municipal General' and renamed the hospital as SWAN MG Hospital. It claimed that such renaming was effected to avoid diminishing SWAN's 'brand value'. In addition, the hospital also transformed its looks, ethos, and practices along commercial lines. With the granted autonomy, SWAN MGH developed super speciality facilities.[28] Within a year, it set up another floor, augmented medical and diagnostic facilities for private patients, launched 'revenue generating' health packages for corporate clients, and developed other specialities like oncology, cardiac surgery, and sports medicine.

In contrast to these improvements in the fee-paying parts of the hospital, services that general category patients have access to have not been expanded. The poor are still entitled to only a small segment of the services provided by the hospital in the four basic specialties of general medicine, general surgery, paediatrics, and obstetrics and gynaecology, and diagnostic facilities in pathology and radiology. For all other services, general patients have to pay private charges. Only recently, a single ICU bed was brought under the 'reserved' category after protests from activists.

---

[28] SWAN MGH now has four major operation theatres (OTs), a cardiac centre with its own OT and cardiac catheterization lab, a 6-bedded neonatal intensive care unit (NICU), a 10-bedded intensive care unit (ICU), a 4-bedded high dependency unit, a 10-bedded artificial kidney dialysis unit (AKDU), CT-Scan, mammography, blood bank (with component facilities), and a sports medicine centre.

1. Profit motive, missing cross-subsidization

The emergence of PPPs in health is an outcome of the notion that greater involvement of markets will improve the availability and quality of health care services for the poor. However, the case of the SWAN MGH PPP shows that this policy has resulted in quite the opposite of what was intended. Rather than improving services for poor patients, it has led them to experiencing new and greater exclusions, inequalities, marginalization, and discrimination. The drive for greater profits has obscured the underlying idea of cross-subsidization of services for the poor. SWAN MGH has been mired in controversy right from the beginning and has faced allegations of non-compliance, restricting access, and fraud. Over the years, a number of petitions have been filed with MCGM regarding the inadequacy of allotted reservations and non-availability of affordable medical services for the poor at the hospital. The management and staff of SWAN MGH have often been accused of following unethical practices and indulging in irregularities. In particular, poor patients regularly report denial of information about free services, lack of transparency regarding the status of bed occupancy, overcharging, denial of free treatment, un-indicated interventions, unscrupulous billing, inadequate bed allocations, restricted range of medical services, and curtailment of access through restrictive policies.

2. Concealment of information

a. SWAN MGH is perceived by the locals as a private hospital. People are largely unaware that it is a PPP and it is required to provide low-cost treatment for patients admitted under the 'reserved' category. The 'Guide' published by the hospital does not mention anything about the mandatory services for the poor. The hospital also makes no visible effort to disseminate this information or create awareness about its free services. Laila Sheikh, an educated resident of the neighbouring Aligarhi *mohalla* slum settlement explains,

> *I didn't know they have reservations for the poor in this hospital. I now know just because you have told me. If I didn't know this, how will a simple illiterate person know this? Because people don't know what they are entitled to, they don't know how to get it.*

b. It has been alleged that the hospital conceals information about the status of bed occupancy in the general wards. With the general lack of transparency about the occupancy of 'reserved' category beds, poor patients are persuaded to register in the paying category. Activist G. Patil says,

> *Poor patients are often denied admissions on the pretext that free beds are not available. We have caught them lying many times.*

3. Profiteering through unethical means

In spite of being managed by a faith-based charitable trust, the actual conduct of SWAN MGH demonstrates a lack of commitment to its social obligations and goes against the spirit of the PPP agreement. There have been several reported cases of overcharging, extortion, denial of care, un-indicated procedures, malpractices, and unethical behaviour.

a. SWAN MGH charges private rates from patients admitted outside the General OPD hours (8 am–12 pm). A grieving father said this,

> *My son Salman got electrocuted (a month ago). We rushed him to the hospital. We had taken our ration card, but it was afternoon and the social worker wasn't there. They said we can't admit the patient in the general ward because the morning OPD hours were over. We had to admit him as a private patient in the economy class. They asked us to deposit INR 25,000 but we could pay only INR 15,000 at that time. My son died in the hospital on the same day. After his death, they gave us a bill of INR 9,000.*

b. Yet another case highlights this issue of denial of care and failure to treat, this despite court rulings that mandate immediate treatment and stabilization of all patients without discrimination during medical emergencies. Ismail Khan, a resident of Aligarhi *mohalla* shared,

> *I took my mother to SWAN Hospital when she was breathless one night. The doctor told us to deposit INR 25,000 for the ICU or go to David Government Hospital if we didn't have money. I didn't have the money then. I told them I'd pay next morning. We offered to keep our jewelry and mobile phones as collateral, but they refused to admit her. They called an ambulance to take her to David Government Hospital. My mother died on the way.*

c. Promoting un-indicated procedures like ultrasound, unnecessary caesarean sections, or incubators for the newborn is common occurrence. The experience of Sayabi Khan of Bluebird Hill illustrates this. She narrates,

> *The doctor (at SWAN MGH) told my daughter-in-law (Sabina) that she would need to undergo a caesarean section. She got frightened. We went to the government maternity home at Rahinagar. The doctor (at Rahinagar) said that everything was okay and an operation was not needed. Sabina delivered normally three days ago.*

d. Other unethical practices such as charging for unused medicines have also surfaced. Yasmeen Sheikh, whose father-in-law expired while admitted at SWAN MGH, gives one such example,

> *We had purchased medicines worth INR 32,000 from their pharmacy just before my father expired. We asked the hospital for a refund for the medicines that were left unused. They said that they had used them all. Then one of the nurses asked us to go and quietly check in the pharmacy. The medicines were kept in a blue bag so we were able to easily recognize them when we saw the bag in the pharmacy. We had a big fight with the hospital staff before they returned the amount to us. We were told by insiders that this was their normal policy.*

4. Restricted access

Hospital policies and guidelines have been instrumental in dissuading poor patients and curtailing access. Rules are framed to help circumvent contractual commitments under the MoU.

a. The hospital does not accept pregnant mothers who register late. Suleiman Gillani of Suhani *chawl* narrates his experience,

> *They did not register my pregnant wife because we went late. They refuse to register or treat any pregnant women who come during or after their third month of pregnancy.*

b. General category patients who report beyond the general OPD hours (8 a.m.–12 p.m.) are not entertained. Sheikh Ibrahim, a resident of the Suhani *chawl* slum settlement, said,

> *I took my son to the hospital for treatment. By the time my turn came, it was already 11 a.m. Their OPD is supposed to be till 12 (noon). However, if you are unable to reach the doctor by 11 a.m. or if they have finished their daily quota, they simply ask you to go away.*

c. A social activist sums up the issue,

The quality of care at this hospital is much better. It has better physical infra-structure, it offers better medical and nursing services, the premises are cleaner and the food is better. But if you hardly have any access to these facilities, then what is the point?

5. Inadequate provisioning of services

a. The conversion of the facility from a maternity home to a general peripheral hospital reduced the number of beds reserved for free maternity care from 86 down to 6–7. Recounting her experience, 24-year-old Sajeeda Khatun of Aligarhi *mohalla* said,

We have to spend the whole day and extra money for travelling to Rahinagar to seek maternity services. My sister was also turned away from SWAN MGH and then the maternity hospital at Rahinagar was also full, so we had to take her all the way to the tertiary public hospital at Sion.

A social activist adds,

Past generations of local residents mostly delivered at this maternity home. This was a well utilized facility, used by working class women and slum dwellers in this area. The generation born here, has now grown up. But they have nowhere to go. Six beds for free maternity care as against the earlier 86 beds are like a drop in the ocean.

b. Other medical facilities and services available to the poor are also inadequate. Irfan Mohammed of Aligarhi *mohalla* highlights how a single 'reserved' ICU bed is a poor joke on needy patients,

When a patient goes to SWAN MGH, they will ask the patient to get admit-ted to the ICU. But they will not talk about the reservation they have for general patients in the ICU. And if the ICU has only one bed reserved, what's the use? How will it ever be enough?

★ ★ ★

The chapter has argued that neo-liberal forms of organizations that have been proactively ushered in by the government under PPP (like the SWAN MGH in Mumbai), have failed to meet their stated policy goals. Instead of improving services for the poor, such initiatives are breeding greater exclusion and marginalization. Findings suggest that PPPs in their current form, as a surrogate to the state's social responsi-bilities (like health care), are ill-advised on many counts.

In this particular case, the state has justified a shift in social policy on the pretext of lack of funds, and in the process neglected its responsibility towards the health of the poor. Under MCGM's market-oriented philosophy, servicing the poor was seen as a drain on the state's resources. The government thus justified the reconfiguration of the public health care apparatus in the name of catering to the poor through market mechanisms, while creating more benefits for paying patients and the private partner. The more damning implication is that even the bureaucracy is not always guided by a culture of public service. Elements within the *babudom* are no less commercial than private players and often facilitate this process in many ways.

Inequity of provisioning is covertly embedded in PPP MoUs through active connivance of private players, government functionaries, and politicians. The private partner disproportionately profiteers through this medical, administrative, political, and business nexus, and gains long-term benefits like commercial use of prime land either free or at nominal cost, tax benefits, subsidies, and considerable freedom to develop and expand revenue-generation activities. While the middle and rich classes utilize these facilities, a gap between the agreement and its implementation ensures shrinking provision, reduced access and unimaginable hardships for the poor in deriving legitimate benefits from these projects. Commitments are not honoured through curtailment of service, blocking access, rude behaviour, and unethical practices, resulting in a steep rise in actual health care costs for the poor. The poorest of the poor go without any health services or receive substandard treatment, resulting in poor quality of outcomes.

Empirical evidence from this study suggests that existing PPP models in the health sector are tantamount to privatization of public assets and the attendant destruction of public health institutions, particularly in the absence of any regulatory framework. In contrast, while public health care facilities may not have been able to attain the desired levels of efficacy, their intent to ameliorate the pain and suffering of the poor has never been in question as their mandate is better defined.

This chapter concludes that there is no ground to either dismantle public health care or place undue faith in the market economy for fulfilling the state's social obligations. This would only push the poor into further deprivation or marginalize them, grievously affecting their health. Along with disenchantment with the ability of PPP-like initiatives to deliver public good, there is a growing conviction that revival and strengthening of the public health system is urgently called for.

Note: This chapter is part of the larger research initiative of IHD–OXFAM India titled 'Study on Public–Private Partnership in Social Sectors'. We are grateful to Sonya Gill for facilitating data collection, Sudipto Patra for editorial work and Professors Padma Velaskar and Imrana Qadeer for comments on the chapter. The chapter also benefited from the comments on the preliminary findings presented at the 13th International Conference on Maharashtra: Culture and Society, Slovak, Bratislava, June 2010.

5 Commercialization and the Poverty of Public Health Services in India

Rama Baru

This chapter examines the transformation of commercial medicine over the last six decades. Much of the scholarly writing on this subject has tended to focus on the characteristics, distribution, quality, and role of the 'for-profit' segment of the private sector.[1] A few scholars have, however, argued that it is important to acknowledge that there is a mixed economy in health services.[2] The public and private sectors have not grown independently of one another. The latter has used the former as a springboard for its growth resulting in the blurring of boundaries between these two sectors. There are several pathways through which these two sectors are interrelated. These include private practice by government doctors, their engagement with the private hospitals as consultants, and the role of commissions in referral for clinical and

[1] Gita Sen, A. Iyer, and A. George, 'Structural Reforms and Health Equity: A Comparison of NSS Surveys, 1986–87 and 1995–96', *Economic and Political Weekly* 37, no. 14 (2002): 1342–52.

[2] T.R. Dilip, 'Extent of Inequity in Access to Health Care Services in India', in *Review of Health Care in India*, ed. V.G. Leena, Ravi Duggal, and Abhay Shukla (Mumbai: Centre for Enquiry into Health and Allied Themes, 2005), 247–68.

diagnostic testing to the private sector. Further, the private sector has grown as a result of various subsidies and incentives provided by the public sector. These include investment in medical, nursing and paramedical education, bank loans for doctor entrepreneurs, the reduction of import duties on high technology medical equipment, granting of an industry status to hospitals to facilitate access to loans, and subsidies for infrastructure from investment companies.

For the purpose of this chapter we choose to employ the concept of commercialization instead of privatization because the former goes beyond the activities of the private sector alone and brings into its ambit the increasingly commercial orientation of publicly owned and non-profit institutions. This concept was used to study the commercialization of health services in low and middle income countries by Mackintosh and Koivusalo (2005).[3] Most health service systems across developed and developing countries have been transformed along the principles of commercialization and India is no exception. Mackintosh and Koivusalo (2005) define commercialization as

> the provision of health care services through market relationships to those able to pay; investment in, and production of those services, and of inputs to them, for cash income or profit, including private contracting and supply to publicly financed health care; and health care finance derived from individual payments and private insurance.[4]

That is, even the public and non-profit sector can become commercialized without getting privatized. This helps us to study the trends going beyond the growth of the private sector, especially private corporate sector, and analyse how unregulated commercial interests change values in medical practice.

This chapter will analyse the process of commercialization and its transformation since independence in India. It identifies and elaborates how the process of commercialization has contributed to the many anomalies that we see at present in the Indian health services. The many systemic weaknesses in the public sector is reflected in the patterns of inequality in access to health services, rising costs, high out of

---

[3] M. Mackintosh and M. Koivusalo, eds, *Commercialization of Health Care: Global and Local Dynamics and Policy Responses 2005* (New York: Palgrave Macmillan, 2005).

[4] Mackintosh and Koivusalo, *Commercialization of Health Care*.

pocket expenditure and lack of satisfaction with the quality of services. This results in people shifting from the public to the private sector, that is largely unregulated, exploitative with differential quality of care. The crisis in the health services exemplifies the fault lines of India's socioeconomic development.

This chapter is divided into the following sections—the history of markets in medical care; the changing relationship of market and state; the characteristics of the private sector and lastly, how markets are able to exert their influence and transform the culture and behaviour of public and not-for-profit services.

THE HISTORY OF MARKETS IN MEDICAL CARE

At the time of independence commercial activity was largely restricted to individual practitioners and the pharmaceutical industry. The Bhore Committee Report in 1946 took cognizance of the fairly large presence of individual private practitioners at the time of independence.[5] According to an estimate the health services were dominated by private, individual practitioners. The Committee was of the view that presence of individual private practitioners was not a threat to the idea of a national health service therefore did not challenge their interests. By choosing not to challenge private interests it implicitly amounted to its accommodation in the Committee's blueprint for health service development. The Committee was of the opinion that private practice should not be allowed by government doctors because this would lead to differential treatment based on the ability to pay. Therefore it would lead to a dual system of care and will undermine the principle of equal access and treatment to all. The rationale that guided the Committee's thinking on the role of the private practitioners was based on the assumption that a strong public sector would make the private sector redundant over time. The committee recognized that large public investments were required in order to build a strong national health service. They had recommended that 12 per cent of the total plan outlay must be devoted to health. However the first two plans earmarked a mere 2–3 per cent for health. Due to the low level of investments,

[5] Bhore Committee, *Report of the Health Survey and Development Committee*, vol. 2 (Delhi: Government of India, 1946).

the public sector was weak and the vision of the Bhore Committee could not be fully realized. Thus the under-investment of the public sector enlarged the spaces for commercial activity within and outside public provisioning.

The Mudaliar Committee's recommendation to employ private practitioners in government hospitals in order to overcome the shortage of doctors was a continuum of the thinking of the Bhore Committee.[6] The latter had suggested this as a possible short term strategy in the context of shortage of trained specialists which was reiterated by the Mudaliar committee report. The Committee was of the opinion that private practitioners should be 'given opportunities to serve in government hospitals on a part-time or honorary basis and the hospital authorities should encourage them to admit their patients needing in-patient care'.[7] This kind of an approach led to the legitimization of private doctors in public hospitals thus creating a public–private mix in the health services.

Thus the accommodation and legitimization of commercial interests in the public sector began as early as the mid 1960s. At this stage the public–private mix was largely restricted to individual private doctors since the proportion of private hospitals was very small.

The relationship between the public and private sector transformed with the growth of private nursing home and hospitals during the 1970s. There was a regional dimension to this growth during this period in the Western and Southern parts of the country. With the growth of the private institutions, it increasingly became that government doctors who were practising privately started engaging with the private sector as consultants. They used their position as government doctors to refer patients to the private sector. This back and forth linkage between the public and private sectors made it difficult to clearly demarcate the boundaries between the two.[8]

[6] Mudaliar Committee, 'Report of the Health Survey and Planning Committee' (New Delhi: Government of India, 1961).

[7] Mudaliar Committee, 'Report of the Health Survey and Planning Committee', 136.

[8] Indian Council for Social Science Research (ICSSR) and Indian Council for Medical Research (ICMR), *Health for All: An Alternative Strategy*, Report of the Study Group set up jointly by ICSSR and ICMR (Pune: Indian Institute of Education, 1981).

CHARACTERISTICS OF THE MIXED ECONOMY IN MEDICAL CARE

Government under-investment in the public sector during the first few decades after independence created a system ridden with anomalies and weaknesses. An example of this is the accommodation of commercial interests, mainly in provisioning and pharmaceuticals. This resulted in a mixed economy in medical care. Much of the writing on the for-profit sector has examined it independently of the public sector that provides only a partial view of the characteristics of the mixed economy in medical care. Commercial interests were first accommodated by the State in the form of private practice by government doctors, doctors from the private sector acting as consultants in government hospitals, creation of private wards, and paying beds in public hospitals.

By the 1970s, there was an expansion of private nursing homes and hospitals which led to the growth of complex inter linkages between the two sectors. Most nursing homes had senior government doctors as consultants on their panels. This resulted in the diversion of patients from government hospitals to the private sector. There were instances when patients would pay to consult a specialist in the government hospital. If they required in-patient care they would be referred to a private nursing home or would make informal payments to the doctors in order receive treatment in the government hospital.[9] Therefore, patients had to pay for care in both the sectors leading to the blurring of boundaries between the public and private sectors. Although efforts were made by state governments to ban private practice, it was often met with opposition from the Indian Medical Association indicating how deep rooted the nexus was between the two. Given the interrelatedness of the public and private sectors, it was increasingly difficult for government to demarcate and address their roles.[10] This was seen in several states at different points in time. For example, during the 1980s, the government of the erstwhile Andhra Pradesh had passed an order banning private practice by

[9] K. Venkatraman, *Medical Sociology in an Indian Setting* (New Delhi: Macmillan, 1973); Rama Baru, *Private Health Care in India: Social Characteristics and Trends* (New Delhi: Sage Publications, 1998); P. Rama Devi, 'Practices of Doctors in Some Government Hospitals in Hyderabad', unpublished Mphil dissertation, 1985, University of Hyderabad.

[10] ICSSR and ICMR, *Health for All*.

government doctors. Soon after, the government doctors went on strike and reversed the ban through a court order.[11]

Increase in the number of nursing homes and diagnostic facilities in the private sector led to intense competition for patient supply. As a result, the role of commissions to doctors in both sectors for prescription of drugs and referrals for diagnostic testing became fairly rampant. The growing influence of the pharmaceutical and technology industry since the 1970s has been fairly well documented. That it is pervasive in influencing clinical decisions of doctors in both the public and private sectors is also well known. The pharmaceutical industry has been employing marketing strategies through their representatives to promote their products and manage to influence the clinical decisions of doctors, overtly and subtly. For example, the role of commissions in the form of cash incentive, sponsorship of conferences and research by the pharmaceutical industry are well known.[12] These are some of the pathways through which commercial interests of the medical industry influence institutions and individuals. Nagral explains the nexus between doctors and diagnostic centres as follows:

> It must be stated at the outset that the practice of giving commissions for referral of patients is not restricted to the GP-specialist interaction. It is now commonplace for commissions to be given by pathology laboratories, radiology establishments, equipment manufacturers and perhaps even institutions. In fact, even specialists practice a sophisticated form of commissions by referring patients to each other, often more as a 'return referral' than because there is a genuine need. Also, it is probably true that the idea of such commissions originated from aggressive specialists trying to increase their practice through commercial incentives.[13]

CHARACTERISTICS OF THE FOR-PROFIT AND NON-PROFIT SECTORS

At present, the for-profit sector is characterized by plurality and heterogeneity in terms of providers and institutions. It can be visualized as

[11] Baru, *Private Health Care in India*.

[12] Chandra Gulhati, 'Marketing of Medicines in India', *British Medical Journal* 328, no. 7443 (2004): 778–9.

[13] Sanjay Nagral, 'General Practice: Some Thoughts', *Issues in Medical Ethics* 10, no. 2 (2002): 15–16.

a pyramidal structure with the primary level at its base which is largely the domain of individual practitioners who are formally and informally trained across systems of medicine. These practitioners are distributed across both rural and urban areas. The secondary level that occupies the middle segment of the pyramid consists of small and medium nursing homes promoted by a single owner or partnerships, mostly among doctors. The spread of these institutions is marked by inter-state and rural–urban variations. The states of Punjab, Gujarat, Maharashtra, Andhra Pradesh, Karnataka, Tamil Nadu, and Kerala are seen to have a higher proportion of 'for-profit' institutions than public institutions compared to other states in India. The reason for these states exhibiting a higher growth could be due to a variety of factors. First, agrarian prosperity has helped to create surpluses that are necessary for investment in a variety of non agrarian entrepreneurial activities. Among a range of these activities, one finds that investment in higher education that includes medicine and engineering is a preferred choice. Medical care, hospitals, diagnostic facilities, and laboratories are also seen as areas of returns for investments. This could explain the growth of private medical colleges in the southern and western regions of the country. Till about the 1970s medical colleges were mostly in the public sector. This was in keeping with the Nehruvian vision of self-reliance and creating the scientific and technical personnel for building a modern India. The sole exceptions to this rule were the few missionary medical colleges like Christian Medical College in Vellore and Ludhiana. The growth of private medical colleges begins in the late 1970s in Karnataka and Maharashtra. Andhra Pradesh, Tamil Nadu, and Kerala follow suit from the 1980s to the present.[14] An analysis of the social background of promoters of private medical and engineering colleges suggest that they belong to the intermediate castes with agrarian roots and political parties.[15] The growth of private medical colleges is part

[14] 'Commercialisation of Medical Education', Presentation, CESI Annual conference on 'Rethinking Education Policy', University of Hyderabad, 16 November 2011.

[15] Sangeeta Kamat, 'Neoliberalism, Urbanism and the Education Economy: Producing Hyderabad as a "Global City"', *Discourse: Studies in the Cultural Politics of Education* 32, no. 2 (2011): 187–202; R. Baru, 'Commercialisation of Medical Education', presented at the CESI Annual conference on Rethinking Education Policy, held at the University of Hyderabad, 16 November 2011;

of a larger process of the movement of regional private capital into higher education, especially engineering, IT, and medicine. These institutions were promoted by the intermediary castes like the Marathas in Maharashtra, Reddys and Kammas in Andhra Pradesh, Chettiars, Mudaliars, and Gounders in Tamil Nadu, and the Patidars in Gujarat.[16]

The data on the establishment of private medical colleges shows a steady growth from the 1970s to the 1990s, with a sudden spurt during the last five years. In terms of distribution of these colleges, over 40 per cent of the private medical institutions are concentrated in Karnataka, Maharashtra, Tamil Nadu, Kerala, and Andhra Pradesh.[17] These colleges admit students on capitation fees. It is estimated that on an average a student pays INR 25–30 lakhs for undergraduate admission and for a post graduate seat, it is over a crore of rupees.[18] Young graduates from these colleges are under financial pressure to recoup the large investments that they have made in medical education.[19] The majority of them tend to look for jobs in the private sector. Some find employment in hospitals, others join the family enterprise and a few invest in hospitals. Marriage is another source for recovering the investments made in medical education. Receiving a high dowry by marrying off a doctor groom is also another way to recover the expenditure incurred

---

A. Diwate, 'Social Characteristics of Private Medical Colleges in India: A Study of Select States', ongoing PhD Work, Centre of Social Medicine and Community Health, Jawaharlal Nehru University, New Delhi, 2013. Also see Nambissan's argument on the role of middle classes in accessing private schools and professional colleges; Geetha Nambissan, 'The Indian Middle Classes and Educational Advantage: Family Strategies and Practices', in *The Routledge International Handbook of Sociology of Education*, ed. M. Apple, S. Ball, and L. Gandin (London: Routledge 2010), 285–95.

[16] Harish Damodaran, *India's New Capitalists: Caste, Business and Industry in a Modern Nation* (London: Palgrave Macmillan, 2008).

[17] Diwate, 'Social Characteristics of Private Medical Colleges in India'.

[18] N. Ananthakrishnan, 'Regulation of Medical Education and Practice', Panel Discussion, National Bioethics Conference, All India Institute of Medical Sciences, New Delhi, 17 November 2010; Ramya Kannan, 'Charging an Arm and a Leg for a Medical Seat', *Hindu*, Bangalore, 28 June 2013.

[19] Ananthakrishnan, 'Regulation of Medical Education and Practice'; Kannan, 'Charging an Arm and a Leg for a Medical Seat'.

on their medical training.[20] Therefore, the growth of private medical colleges has abetted commercialization of medical care and has facilitated the inter-generational transfer and consolidation of ownership of private enterprises.

CHARACTERISTICS OF NON-PROFIT MEDICAL CARE

Similar to the for-profit medical care institutions, the non-profit medical care sector is also characterized by plurality and heterogeneity. The dominant institutional forms in the non profit sector include charitable dispensaries and hospitals. Most of these hospitals offer a range of general services with a few that provide mainly specialist services. There is heterogeneity in the promoters of these institutions. These include faith-based organizations, traders, or industrialists before and soon after independence.[21] For greater analytical clarity, it is useful to draw a distinction between the old trust and new trust dispensaries and hospitals. The old trust, dispensaries and hospitals, were set up during the latter half of the nineteenth and early twentieth century, while the new trusts are those that came up in the latter half of the twentieth century.

By the 1980s new medical care trusts began to appear. They were quite distinct from older health care trusts. The new medical care trusts were promoted by big business groups by setting up super speciality hospitals.[22] 'New' trust hospitals were able to assume trust status under the claim of having a research wing in the hospital. This provided the rationale to access public subsidies in the form of tax exemptions and land at subsidized rate.[23] What one is arguing here is that the motivation for assuming the status of a trust was guided by the goal of profit maximization. It is well known that medical care requires human and

[20] M. Rao, K. Rao, A.K. Shiva Kumar, M. Chatterjee, and T. Sundararaman, 'Human Resources for Health in India', *The Lancet* 377, no. 9765 (2011): 587–98.

[21] Baru, *Private Health Care in India*; Damodaran, *India's New Capitalists*; M. Nundy, 'Social Transformation of "Not-for-profit" Hospitals in Delhi', PhD dissertation, Centre of Social Medicine and Community Health, Jawaharlal Nehru University, 2010.

[22] Nundy, 'Social Transformation of "Not-for-profit" Hospitals in Delhi'.

[23] Nundy, 'Social Transformation of "Not-for-profit" Hospitals in Delhi'.

technical investment for a long period before it becomes financially viable and profitable. Therefore by acquiring the status of a trust they were able to become financially viable in a relatively shorter period of time. Although research was the reason for getting the status of a trust, there is very little information regarding the academic output of the institutions that are registered under this clause. In fact it is very hard to distinguish between the trust and corporate hospitals in terms of the services offered, costs, appearance, and the general ethos.

RISE OF CORPORATE MEDICAL CARE

The most recent and significant transformation is the rise of corporate hospitals in India. The first of its kind was the Apollo Hospitals that was promoted by Dr Prathap Reddy in the late 1980s in Chennai. Dr Prathap Reddy whose family belongs to the erstwhile Andhra Pradesh, practised in the US for many years. In 1979, he had visioned a corporate model for health care in India. Although he did not have a business background in running a corporate hospital, he was the first to break the taboo of profit making in health care. As a non-resident Indian (NRI) doctor he was influenced by the American model of medical care. In 1983, he set up the first Apollo hospital with seed money of INR 1.3 crores that he mobilized from his family and partly from doctors in India and abroad. The groundwork included getting access to finances, land and concessions for import of medical equipment. All this was possible through active lobbying with the political class. Dr Reddy met the then Prime Minister, Mrs Indira Gandhi, who facilitated the healthcare sector acquiring the status of an industry and gaining access to financial markets. Subsequently, during Rajiv Gandhi's tenure as Prime Minister, Apollo Hospitals was able to set up a joint venture with the Delhi government, the Indraprastha Apollo, the first-ever joint venture between a state government body and a hospital. Following the suit of Apollo hospitals, Max and Fortis also entered the market. Interestingly, Fortis Hospitals is registered as a Trust and is not technically a corporate hospital. It behaves very much like a corporate hospital in terms of infrastructure, cost with a focus on specialist services. The promoters of both Max and Fortis were part of the family that owned Ranbaxy, a major pharmaceutical company in India. In Chennai, Malar Hospitals and Devaki Hospitals, among others, were promoted as corporate concerns. A common thread that connects these various enterprises is the link with the NRI doctors who chose to return to India either as promoters or work as professionals in these hospitals.

The corporate enterprises have themselves transformed over the last decade. These changes have transformed the landscape of health service provisioning. Initially, the corporate sector was associated with tertiary specialist services. However, from 2007 onwards they have diversified their operations in terms of operations and service delivery. In terms of operations there is plurality in the institutional arrangements. These include institutions promoted by the corporate entity alone, joint venture between the corporate enterprise with the government and joint venture between the corporate enterprises with smaller private hospitals.

In terms of service delivery there is a move to build a network of institutions at the primary, secondary, and tertiary levels. The three levels are now vertically integrated, similar to the public sector. In the process they are able to provide services at all levels of care, both preventive and curative. Given the weaknesses inherent in the public health services delivery, they are able to potentially source patients at the primary level. This becomes a means to ensure a steady supply of patients and since they are vertically integrated, they are serviced by the same organization at all levels. This structure of service delivery helps to rationalize cost because the bulk of out-patient care is for routine illnesses that occur much more frequently in a population as compared to those requiring high-end care. Some of the corporate enterprises have integrated diagnostic centres and pharmacies with the institutions at the different levels.

COMMERCIALIZATION AND
THE NON-PROFIT SECTOR

With the expansion of the 'for profit' sector since the 1970s, the spaces for the old 'non-profit' institutions have started shrinking. This is seen especially in the case of the hospitals under the Christian Medical Association of India for which data is available. The data suggests that during the 1940s the Christian Medical Association had the largest number of hospitals and medical colleges. By the 1970s, several of these were forced to close down.[24] A complex set of reasons seem to be responsible. These include the economic recession, the reduced flow in funds from church donors, the difficulty in paying salaries to doctors that matched the amount offered by public and for-profit institutions; the rising cost of technology, and debates regarding the appropriate deployment of technology.[25] In order to bridge the revenue gap due to increasing cost of care, several mission hospitals introduced user charges. There was a division among the missionary institutions regarding this issue. Those who chose not to introduce user charges were forced to close down. Those who went ahead with it faced the moral dilemma of excluding those who needed care the most because of their inability to pay.

Apart from the missionaries, during the late nineteenth and early twentieth century, the Marwaris invested in non-profit dispensaries and hospitals, mostly in Rajasthan and West Bengal.[26] Often small with fewer than 30 beds in size, these institutions provided general services. These institutions received public subsidies, mostly in the form of land, during the post-independence period and several of these prided themselves in providing services to the poor and needy.[27]

With increased competition from the 'for-profit' sector, these institutions were forced to either close down or look for large investments to finance infrastructure and technology. A study of the transformation of non-profit hospitals in Delhi shows the emergence of partnerships between non-profit hospitals and corporate entities.[28] The façade of the trust is used for getting tax exemptions while the corporate bodies invest, expand, and manage these hospitals.[29]

---

[24] Baru, *Private Health Care in India.*

[25] Baru, *Private Health Care in India.*

[26] Madhurima Nundy, *The Not-For-Profit Sector In Medical Care: Financing and Delivery of Health Care Services in India*, background papers, National Commission on Macroeconomics and Health, New Delhi, Ministry of Health and Family Welfare, Government of India, 2005.

[27] Rama Baru, 'Private and Voluntary Health Services: An Analysis if Inter-Regional Variations', 1996, project report submitted to UNDP; Nundy, *The Not-For-Profit Sector in Medical Care.*

[28] Nundy, 'Social Transformation of "Not-for-profit" Hospitals in Delhi'.

[29] Nundy, 'Social Transformation of "Not-for-profit" Hospitals in Delhi'.

TRANSFORMATION OF 'FOR-PROFIT' MEDICAL CARE

Until the early 1970s the for-profit sector was largely restricted to individual practitioners. Subsequent to this there have been several waves of transformation in the for-profit sector. The first wave of transformation of this sector was witnessed during the mid-1970s. During this period there was a growth of small and medium sized nursing homes that were promoted by doctor entrepreneurs. Majority of these doctor entrepreneurs belonged to the upper caste/classes from urban and rural areas.

This was partly facilitated by the investment made by rich and middle peasants in higher education that gave rise to a professional class that had a rural base. This was particularly striking in states that had capitalized from the introduction of the green revolution technology during the 1960s. The surpluses accrued from the green revolution also fuelled investments into a range of commercial activity that included private nursing homes and hospitals.[30] A study of inter-regional variations in the growth of health services in the Telengana and Coastal Andhra regions of erstwhile Andhra Pradesh shows that the number of private institutions is almost double of the former. This is because of the differences in the history and economic transformations between these two regions.[31]

The second wave of transformation was the early 1980s when the health policy document states that the government is not in a position to meet the rising demand for health services and therefore sought the co-operation of the for-profit and non-profit sectors.[32] This statement is important because it is for the first time that there is an explicit reference to the role of the private sector in a government document. This provided the legitimacy for the expansion of the private sector with public subsidies at the central level.

[30] D. Thorner, 'Capitalist Farming in India', *Economic and Political Weekly* 52, no. 4 (1969): A211–A212; Gail Omvedt, 'Capitalist Agriculture and Rural Classes in India', *Economic and Political Weekly* 52, no. 16 (1981): A140–A159; C. Upadhya, 'The Farmer-Capitalists of Coastal Andhra Pradesh', *Economic and Political Weekly* 23, no. 27 (1988): 1376–82.

[31] Rama Baru, 'Inter-Regional Variations in Health Services in Andhra Pradesh', *Economic and Political Weekly* 28, no. 20 (1993): 963–7.

[32] Government of India, 'Statement on National Health Policy' (New Delhi: Ministry of Health and Family Planning, 1982).

The third wave of transformation was during the late 1980s, a decade before India embarked on the path of economic liberalization. During the 1980s, government policy was facilitating the expansion of market in the economy and social sectors. In the medical sector, NRI doctors like Prathap Reddy, Naresh Trehan, and others were in dialogue with the then prime ministers Indira Gandhi, Rajiv Gandhi, and later P.V. Narasimha Rao to promote corporate hospitals. Some of the important concessions that facilitated this process were the reduction of government subsidies for import of high technology medical equipment and giving hospitals the status of an industry. These two concessions facilitated and paved the way for the rise of corporate medicine in India. The rise of corporate medicine fundamentally transformed the landscape of the private sector. It resulted in a highly stratified private sector where access to capital defined the scale, scope and power of organizations to influence and define policy.

In my view the rise of corporate hospitals changed the culture of medical practice in several significant ways. First, it raised the cost of medical treatment which was clearly related to the increase in the scale of production and operation. A study of the cost of treatment of specific interventions in the private sector showed a clear differential between corporate hospitals and trust hospitals. On an average, the cost of treatment was almost five times higher in a corporate hospital as compared to a family owned or a trust hospital.[33] Large scale production required investments in infrastructure, human resources, and medical equipment. In order to recover these large investments the culture of medical practice also underwent changes. Government doctors who have moved to corporate hospitals report that the success of a doctor is judged by the number of patients he/she could treat. The volume of patient turnover and diagnostic testing were important for earning profits. As a result there is considerable pressure on the doctors to treat more patients by prescribing diagnostic testing.

[33] Rama Baru, R. Duruvasula, B. Purohit, D. Kumar, and B. Rajesh, 'Efficacy of Private Hospitals and the Central Government Health Scheme: Study of Hyderabad and Chennai', report submitted to the Ministry of Health and Family Welfare, New Delhi, by Administrative Staff College of India, Hyderabad, 1999.

COMMERCIALIZATION AND THE POVERTY OF PUBLIC SERVICES

The process of commercialization has far reaching implications for availability, accessibility, affordability, and acceptability of health services. Evidence shows that it has produced inequalities in access to preventive and curative services. These inequalities when analysed across the axes of caste, class, gender, and region raise some serious concerns for those who have the greatest need but are being excluded because of lack of physical and financial access. The NSSO data on utilization of health services shows a growing dependence on the private sector at all levels for out-patient and in-patient care. The major reason for non-utilization of services, in the public and private sectors is the inability to pay for care. This was largely due to the acceptance of the World Bank policies for health sector reform by the Indian government. This resulted in the introduction of market principles in public services and public subsidies for growth of the 'for-profit' sector at the secondary and tertiary levels. Informed by the principles of new public management, institutional arrangements like public–private partnerships (PPPs) were instituted in the health services. This arrangement further blurred the boundaries and roles between the public and private sectors.

As a consequence of the health sector reforms of the 1990s, market principles in the form of user fees, contracting out of diagnostic facilities, and a variety of PPPs resulted in public institutions behaving like private ones. This resulted in large sections of the middle and working classes exiting public provisioning to the private sector. Studies have shown that a variety of other reasons like lack of availability of health personnel and drugs in the public sector also made people lose trust in its functioning. In addition, the poor interactive quality in public institutions in the form of rude, indifferent and callous behaviour of the staff, created a further distance between the providers and users. At a deeper level, the process of commercialization has altered the aspirations and values of professionals in the public sector.[34] The yardstick

[34] Rama Baru, A. Acharya, S. Acharya, A.K. Shiva Kumar, and K. Nagaraj, 'Inequities in Access to Health Services in India: Caste, Class and Region', *Economic and Political Weekly* 45 no. 38 (2010): 49–58.

of comparison is the corporate sector that has higher salaries, access to high technology equipment, and lesser patient load than public hospitals. This leads to a great deal of frustration among doctors in the public sector that has far reaching consequences on their behaviour as providers of care.

INFLUENCE OF MARKETS ON THE BEHAVIOUR OF PUBLIC AND NON–PROFIT INSTITUTIONS

Rampant commercialization has influenced the structure and culture of public and non-profit institutions at the organizational and individual level.[35] These processes are not adequately acknowledged in the ongoing debate on commercialization. The culture of institutions and the value it embodies is shaped by the larger socioeconomic forces. At a fundamental level, commercialization has led to the commodification of medical care. A retired doctor from a reputed national medical institute observed:

> I was not motivated by money. I think this was true for most doctors of my generation. Probably, this was an individual characteristic which was seen in most other doctors at the institute. I was motivated by the desire to gain name and fame. Being the premiere institute of the country, the most of difficult cases, the most variety of cases came there, which you won't find anywhere else in any other hospital. So, you had the opportunity to continue to learn and to grow…[36]

However, the growth and transformations in the for-profit sector in general and the corporate sector in particular, had far reaching impact on doctors in the public sector. The rise of the corporate sector emerged as a new hegemonic form that became a symbol of world class, quality care. It, therefore, became the standard of good quality

---

[35] Rama Baru, 'Commercialisation and Public Sector in India: Implications for Values and Aspirations', in *Commercialization of Health Care: Global and Local Dynamics and Policy Responses 2005*, ed. M. Mackintosh and M. Koivusalo (New York: Palgrave Macmillan 2005); Nundy, 'Social Transformation of "Not-for-profit" Hospitals in Delhi'.

[36] Interview with a retired doctor, All India Institute of Medical Sciences, New Delhi, cited in Baru, 'Commercialisation and Public Sector in India', 106.

care with respect to which all other forms of institutions were judged. A senior doctor from the public sector opined:

> The growth of the tertiary private sector produced stark differences in working conditions, patient load and salaries as compared to the public sector. The higher salaries in the private sector were attractive. For some, money was an issue because of changing lifestyles, and for others frustrations arising out of lack of promotional avenues and recognition provided the context for questioning and even undervaluing the public sector.[37]

However, the aspirations and values of doctors cannot be seen in isolation of their middle class moorings to which majority belong. Therefore, the changing values of middle class India, of which majority of the doctors were a part, played an important role in shaping professional values.

CHANGING VALUES OF MIDDLE CLASS INDIA AND SHAPING OF PROFESSIONAL VALUES

Several scholars have studied and commented on the rise of the new middle class during the 1980s as distinct from the old middle class in terms of values.[38] They rightly argue that the old middle class did not indulge in conspicuous consumption and encouraged in some cases austerity and simplicity as virtues. However, the period of liberalization with the rise of private enterprise, encouraged and celebrated consumerism. This has led to changes in lifestyle and the redefinition of status, aspirations, and values of the middle class. As a retired doctor observed:

> The coming in of globalization and consumerism...consumerism and the desire and the availability to people also added a materialistic approach. Society is wanting more than what they were. Doctors are not left behind. Their salary structure [in the public sector] had not improved with time... Lifestyles of people in the private sector appeared to be better. Lifestyle differences became glaring. The younger

---

[37] Baru, 'Commercialisation and Public Sector in India', 106, 110.

[38] B. Mishra, *The Indian Middle Classes: Their Growth in Modern Times* (Delhi: Oxford University Press, 1961); L. Fernandes, *India's New Middle Class: Democratic Politics in an Era of Economic Reform* (Delhi: Oxford University Press, 2007).

people found their counterparts in the business world, especially MBAs, making huge amounts of money. That's when they thought to themselves, 'Why can't I afford and live their lifestyles?'... This produced dissatisfaction and frustration with the public sector.[39]

Commercialization of medical care has changed the culture of medical practice. Corporate hospitals have set the trend in access to high end technology and doctors tend to rely more on diagnostic testing to arrive at a diagnosis. Those who seek care are also influenced by these trends and equate good quality medical care with high technology.

★ ★ ★

This chapter has tried to examine the transformation of commercial medicine over the last six decades in India. It argues that the private sector has been dependent on the public sector for its growth. Over the period, the private sector has become much more assertive and influential in shaping health policy at the state and national level. In more recent times, there has been a consolidation and assertion of corporate interests—hospitals, insurance, and pharmaceuticals—on public policy. This clearly paves the rise of the medical industrial complex in India modelled along the lines of American medical care.

[39] Interview, cited in Baru, 'Commercialisation and Public Sector in India', 110–11.

6 'It All Changed after Apollo' and Other Corporate Hospital Myths*

Sarah Hodges

INTRODUCTION

In the shadow of recent proposals for Universal Health Care (UHC) in India, discussions regarding the impact of private medical care on Indians' health have taken on a greater urgency. However, due in large part to an endlessly delayed Clinical Establishments Act (and delays to its implementation), our collective attempts to evaluate the effects of India's growing private medical sector have been seriously hampered. This is due to a lack of reliable or comprehensive data regarding (*a*) the size of the private health care sector and (*b*) its patterns of growth, particularly since the 1980s. As we formulate and assess practical strategies for a sustainable health care future for India in the absence of reliable statistical data, historians' tools for understanding the recent career of health care in India merit consideration. This is because these tools—in particular, the internet-as-archive of official and unofficial documentation, as well as oral history interviews—allow us to comprehend and interrogate a set of 'common-sense' understandings of the recent past of health care in India. Attending to our assumptions about the past

* A different version of this chapter was published under the title '"It All Changed after Apollo": Healthcare Myths and Their Making in Contemporary India' in the *Indian Journal of Medical Ethics* 10, no. 4 (2013): 242–9.

matters, because our understanding of the past grounds our ability to imagine a different future.

To that end, in this chapter, I use material gathered from interviews (conducted in 2009–10) with twenty prominent physicians in Chennai as well as materials gathered from newspapers and official and unofficial health-related documentation to examine a touchstone in India's recent health care history: that of the 1983 launch and subsequent three-decade career of Apollo Hospitals in Chennai. I argue that one of the singular successes of Apollo Hospitals (and its founder and Chairman, Prathap Reddy) has been in image management. Because I am particularly concerned with the construction and circulation of myths surrounding Apollo Hospitals, I explore the myths that surround Apollo and Reddy, and explore the corollary phenomenon of myth-making. In particular I investigate not only the myth-making activities of Apollo and Reddy, but also how these stories are regularly reproduced within a wider (and often, ironically, a critically engaged) community of medical professionals across the region. To anticipate the argument somewhat, the Apollo 'success story', while central to today's dominant narrative of the economic growth produced by India's private health care sector, is based on assumptions and assertions. These assumptions and assertions begin to crumble even under the most basic historical scrutiny. In short, in its three decades, Apollo's greatest success may perhaps be its story.

We are now lucky to have at our fingertips a growing field of the work that documents and analyses the history of health care in modern India. For the most part, historians have told the story of health care in India as part of state practice. This history begins with analyses of significant policy directives in colonial India, and is followed by similar accounts for independent India.[1] More recently, other historians have elaborated upon these analyses of medicine and power by investigating the relationship between health and development among postcolonial states and within international organizations.[2]

---

[1] David Arnold, *Colonizing the Body: State Medicine and Epidemic Disease in Colonial India* (Berkeley and Los Angeles: University of California Press, 1993); Roger Jeffery, *The Politics of Health in India* (Berkeley and Los Angeles: University of California Press, 1988).

[2] Sunil Amrith, *Decolonising International Health: India and Southeast Asia, 1930–1965* (Houndmills: Palgrave, 2006); Sunil Amrith, 'The Political Culture

However, our unfolding historical moment is now out of step with the analytic frameworks that have grown up around the analytic challenges of, first, colonial health and second, health care and development. The new storyline of India as an 'emergent global power', is one in which India is now a net aid donor rather than recipient. As I have argued elsewhere, whereas India used to represent the past, now it is the future.[3] Today, whether in health care or in other regimes of what they called in the twentieth century, 'welfare', India's greatest challenge is to look after her own. And here the story changes. In particular, in the unfolding history of health care, we now witness the promise and the challenge of what has been called 'Universal Health Care' proposals in India. What was once the state's developmental promise to its people has morphed into a consumer right. In light of this, our current moment demands that we attempt to bring this historiography into dialogue with the new unfolding landscape of health and health care in India.

First and foremost this demands that we must ask a fundamentally historical question, one I first heard posed by Imrana Qadeer in March 2012 at a meeting to discuss UHC. In her characteristically horripilating and take-no-prisoners fashion, Dr Qadeer looked squarely into the eyes of those sitting around the room and asked: 'Why. Health. Now?' Eventually she broke the heavy silence: 'For it's not as though they really care.' She went on to develop a broader argument about how what had been a developmentalist state project of looking after the people's health has now been refashioned into a state making ill-health into a more efficient revenue stream. This has profound implications for the story that we as scholars tell of the recent unfolding of the history of health care as a national project in India. In particular, the history can no longer remain solely a matter of national policy and state institutions. Similarly, its critics (or supporters) can no longer be written off as the representatives of an inefficient state. For Qadeer points to a very efficient state repurposing of health indeed. In light of this, I seek to join other critics and reframe the unfolding of India's recent health care history as one of state policy but one of state policy

---

of Public Health in India: A Historical Perspective', *Economic and Political Weekly* 42, no. 2 (2007); Pratik Chakrabarti, '"Signs of the Times": Medicine and Nationhood in British India', *Osiris* 24 (2009): 188–211.

[3] Sarah Hodges, 'Medical Garbage and the Making of Neoliberalism in India', *Economic and Political Weekly* 48, no. 48 (2013): 112–19.

governing private enterprise.[4] In other words, the recent history of health care in India is as much as anything, a history of the 'corporates'.

APOLLO AND THE INVENTION OF THE 'CORPORATE HOSPITAL' IN INDIA

At the risk of stating the obvious, the rise and spread of Apollo Hospitals from the 1980s matters because it is widely seen to represent the beginning of a new chapter in the history of health care in India: the rise of the corporate hospital. In practical terms, the 'corporate hospital' in India differed from that which came before. These differences were both in terms of the scale and style of health care delivery, as well as in terms of how new hospitals were financed.

Private health care had always comprised a sizable proportion of medical services on offer in twentieth-century India.[5] However, before the invention of the corporate hospital, although charitable trusts could and did run some large hospitals across India, the scale of private institutions was smaller. Private 'nursing homes' were, at their largest, 30-bed institutions. Further, corporate hospitals' promoters linked the idea of large-scale high-quality health care with large, expensive, and often imported pieces of medical equipment. Finally, because the corporate hospital is large (usually having at least 150 beds) and highly technologized, it not only required more money to open but it also heralded on new forms of financing private health care. In particular, in advance of opening hospitals, corporate hospitals' promoters incorporated private companies and sold shares to the public.

---

[4] Rama Baru, 'Privatisation and Corporatisation', in *Unhealthy Trends: A Symposium on the State of our Public Health System*, Seminar 489, Seminar Web Edition, May 2000, 26 March 2012, http://www.india-seminar. com/2000/489/489%20baru.htm; Ravi Duggal, 'Where Are We Today?' in *Unhealthy Trends*, http://www.india-seminar.com/2000/489/489%20duggal.htm; Ravi Duggal, 'Tracing Privatisation of Healthcare in India', *Express Healthcare Management*, 1–15 April 2004, http://www.researchgate.net/profile/Ravi_Duggal/publication/236888943_Tracing_Privatisation_of_Healthcare_in_India/links/0c960519f2e3d3bba1000000.pdf (accessed 8 February 2010).

[5] Rama Baru, *Private Health in India: Social Characteristics and Trends* (New Delhi: Sage Publications, 1998).

Given Apollo's pride of place in 1983 at the heralding of new, post-liberalization chapter of corporate health care in India, Apollo's story has been told and retold among physicians, journalists, politicians, and government bureaucrats and members of the general public. In this story, the career of Apollo Hospitals appears as nothing short of miraculous: Apollo figures simultaneously both as cause *and* effect of India's recent economic successes. Yet, as this essay argues, this legendary status depends on a corollary set of assertions that we may productively refer to as 'myth' because they obscure a set of broader historical processes that both precede and exceed any results that can be attributed to one man or one hospital. In light of this, it is useful to begin with a chronological account, both of the early history of opening of Apollo (particularly in Chennai) as well as an account of the broader regional and national health care provision context from which Apollo emerged.

In 1980, Reddy announced that he had acquired the plot of land for the first Apollo: Apollo Hospitals Chennai, on the centrally located Greams Road. At the time, the *Times of India* reported:

> Dr Prathap C. Reddy, a practising cardiologist and the moving spirit behind the enterprise, told a press conference last night that the Government of India had sanctioned the Rs 7.9 crore project. This would be the first Indian hospital run as a public limited company.... According to plans, the hospital should be ready in January 1982.

The article went on to explain that this was the first of what was a planned chain of hospitals, incorporated as the Indian Hospital Corporation.[6]

This development was significant not only because it was the announcement of the first private limited hospital in India, but also because in order to move ahead with the financing, Reddy had been given permission to build a private hospital with over 30 beds. Up till then, by law only government or charitable trust hospitals were permitted to expand beyond a capacity of 30 beds (Chennai doctor 2[7]).

By 1982, the Reserve Bank of India had given Reddy the permission to secure additional financing for the hospital through a loan of Swiss francs.[8] Further, in this year, Reddy's Indian Hospitals Corporation received permission to offer public issue of shares to complete the financing to pay for the opening of the hospital.[9] This was remarked at the time to be 'the first [public offering] of its kind in India for financing a multi-speciality medical centre to be run on corporate lines'.[10] Rights issuing continued to fund the operation of this and subsequent Apollo Hospitals through the 1980s, 1990s, and beyond.[11]

Amidst much fanfare, in September 1983, Zail Singh, President of India, inaugurated the Chennai hospital. Yet it was not until February 1984, that this Apollo started admitting patients.[12] The hospital consolidated its reputation as the hospital of choice for the city's powerful a few months later, in October 1984, when M.G. Ramachandran (henceforth MGR), the then Chief Minister of Tamil Nadu, was admitted to Apollo.[13] MGR was only the first of many subsequent high profile politician patients who sought treatment in Apollo and whose treatment the hospital publicized widely.[14] This was particularly the

[8] *Times of India*, 'Maiden Public Issue for Hospital Project', Delhi, 17 August 1982.

[9] *Times of India*, 'Consent for New Capital Issues', Delhi, 12 October 1982.

[10] *Times of India*, 'Apollo Hospitals', Delhi, 12 November 1982.

[11] 'The Financial Chronicle Highlights Apollo Hospitals as the Greek God in Indian Healthcare', *Apollo Hospitals Enterprise Limited, 1 November 2011*, 16 March 2013, http://www.apollohospitals.com/media-features-detail.php?newsid=2

[12] *Times of India*, 'Concept of Corporate Hospital Catches On: City Notes', Delhi, 14 December 1984.

[13] *Times of India*, 'MGR in Hospital after Asthmatic Attack', Delhi, 7 October 1984.

[14] Indeed this policy was not always a success. Apollo authorities were charged with a breach of medical ethics in respect of the VIP medical care when K.R. Narayanan, the then Vice-President, was admitted to Apollo for heart treatment:

A section of the medical community here expressed misgivings at the alleged attempt by the Apollo Hospital to publicise Mr Narayana's treatment in a manner that projected the hospital image and professional capabilities of its surgeons.... Following surgery, the head of the hospital appeared on Doordarshan to explain (with a diagram) how the critical blocks in the vice-president's heart vessels were identified and what surgical procedures were adopted by the medical team. (*Times of India*, 'Medical Ethics Violation: Madras Hospital Faces Charge', Delhi, 23 October 1993)

case because beginning in the mid-1980s, Apollo came to be known primarily for cardiac surgery, orthopaedic surgery, and nephrology (Chennai doctor 5[15]). By 1988, Apollo Hospitals expanded to Hyderabad, as Reddy had initially planned. Similarly, in 1996, Apollo Hospitals opened Apollo Indraprastha in New Delhi. Unlike with first Apollo in Chennai, for the Hyderabad and New Delhi hospitals, Apollo Hospitals took out large advertisements in the national press to publicize not only the new hospital but particularly the rights issue.

By 1993, Apollo Hospitals began to take out large advertisements in the national press to congratulate themselves in serving the nation. At the time of writing in 2013, Apollo has undertaken continued expansion in India and beyond. The year 2013 also sees three decades of Apollo to be celebrated by the Group with commemorative volumes by and about Reddy. As one of these promotional pieces recently summed up:

> From one multispeciality facility that he founded in Chennai 30 years ago to 54 hospitals, 1,600 pharmacies, 60 diagnostic clinics and 11 nursing colleges in 2013, Dr Reddy's medical system attracts more than 10,000 footfalls daily across India. Cumulatively, more than 32 million people have been treated at various Apollo hospitals....[16]

APOLLO: MEDICAL MYTH AND MYTH-MAKING

Despite a wide range of opinions regarding the rise of the corporate hospital in post-liberalization India, there is a great deal of convergence in the ways both supporters and critics of Apollo describe its significance. As one admirer recently wrote:

> In 1983, at a time when the government's commitment to investing in public healthcare appeared to be flagging, Prathap Chandra Reddy did something unthinkable: He launched the country's first corporate medical system. Three decades on, the argument over the pros and cons of privatised healthcare in a poor country remains unsettled but there is one thing Dr Reddy's admirers and critics both agree on: the

---

[15] Interviewed on 9 July 2010.

[16] Pranay Gupte, 'Transforming India's Health-care Landscape', *Hindu*, Banglore, 5 February 2013.

emergence and rise of his company, Apollo Hospitals Enterprises, has altered the health-care landscape of India.[17]

Indeed, a critic of Apollo largely agreed:

> I would say that what I noticed during the past thirty years, which is the time I have been practicing medicine, the big change is that when we were undergraduates, there were no private *hospitals* in Chennai. There were private *nursing homes* which was [a] big difference. Because nursing homes wouldn't take acutely ill patients. They would only take elective surgical procedures; very mild illnesses. Anything serious was referred to the government teaching hospitals. Obviously the three: Kilpauk Medical College, Stanley Medical College, and the biggest, Government General Hospital. If you had a serious illness, [in] those days it was considered that the place to go to was Government General Hospital. It all changed after Apollo (Chennai doctor 1).

Yet the claim that 'Apollo changed everything' fails to bear the weight of scrutiny of the broader context of Chennai, in particular, its private health care infrastructure and the career of the Indian government's economic policies and programmes in the 1970s and 1980s. It also does not stand up when viewed against the histories of other large private hospitals—both in Chennai as well as elsewhere across India. It is useful to disaggregate the 'It all changed after Apollo' myth into its five key elements:

1. Apollo came up at a time that health care for 'ordinary Indians' was flagging.
2. Apollo provided a new model of health care delivery in India.
3. At its heart, Apollo is a patriotic project.
4. In order to open Apollo, its Chairman, Prathap Reddy, single-handedly changed government policy.
5. Apollo was an immediate success.

In light of this descriptive convergence among both admirers and critics, the point of this section is to describe and assess these elements of 'It all changed after Apollo' myth.

---

[17] Gupte, 'Transforming India's Health-care Landscape'.

1. **Apollo came up at a time that health care for 'ordinary Indians' was flagging**

One doctor whom I interviewed claimed of Reddy: 'When he set up Apollo Hospitals in Chennai in 1983, private health care institutions were virtually unknown to the country.'[18] This aspect of the myth of Apollo is often articulated through three sub-claims: (*a*) that there was no reasonable health care available in Chennai, (*b*) that the government in particular had either abdicated or was simply unable to fulfil its responsibility to provide health care for ordinary Indians, and that, therefore, (*c*) only the very rich had access to high-quality health care, which they did through foreign travel.

Let us consider these in turn:

1a. *There was no good health care available in Chennai for ordinary Indians*

Although this claim is oft-repeated, it is difficult to find evidence to substantiate it. Indeed, the weight of evidence would point to the opposite: that rather than being a city under-served by high-quality health care, Chennai has been a long-standing centre for medical excellence, far in advance of the opening of Apollo Hospitals in 1983/1984. While scholars have yet to document the city's medical history fully, in my interviews during 2009–10 with twenty of Chennai's prominent physicians, these doctors regularly provided extensive descriptions of Chennai as the long-standing home of high-quality medical care and medical education in India. Nevertheless, none of this is apparent in the following assessment of Reddy and Apollo:

> [Reddy's] plan for the creation of a nation-wide hospital system in the corporate sector may not seem extraordinary today when private medicine has made major inroads across the country but it was dramatic 30 years ago... When he set up Apollo Hospitals in Chennai in 1983, private healthcare institutions were virtually unknown in the country.[19]

On the contrary, the phrase that both physicians in interview and newspaper journalists repeatedly used to describe the cluster of

[18] 'Interview with Prathap Reddy, Apollo Hospitals', *Indian Medicos* (blog), 6 October 2010, http://indian-medicos.blogspot.co.uk/2010/10/interview-with-prathap-reddy-apollo.html (accessed 16 February 2013).

[19] Gupte, 'Transforming India's Health-care Landscape'.

medical excellence in Chennai is 'medical mecca'. Their observations are grounded in detailed discussions of the various aspects of long-standing medical excellence in the city, including: medical education and large government hospitals; a cluster of prominent physicians' private nursing homes (particularly along Chennai's Poonamallee High Road); excellent connections to national transport infrastructure; nodes of specialist expertise; and reputation of ethical treatment available for reasonable fees.

The medical education provided in and around Chennai is legion. Three medical colleges in the region are consistently ranked in the top ten nationally: Christian Medical College in Vellore (established 1902 and affiliated to Madras University in 1942), Jawaharlal Institute of Postgraduate Medical Education and Research in Pondicherry (established 1823 and redeveloped in 1956), and Madras Medical College in Chennai (established 1850). These institutions' students not only staff large teaching hospitals, their graduates also go on to staff and manage the small, medium, and large hospitals across Chennai. Indeed, one doctor whom I interviewed, who was closely associated with the Tamil Nadu branch of the Indian Medical Association, estimated that the city was home to approximately 400 hospitals and 10,000 beds. Further, Chennai is famous for the high quality of treatment and research carried on in a number of its government hospitals, in particular: Government General Hospital (established 1664), Stanley Hospital (established 1792), as well as in its numerous high-profile voluntary and charitable trust hospitals: the Voluntary Health Society (established 1958), the Cancer Institute (established 1954), and Sankara Nethralaya (established 1978). As one physician explained:

> When I went to medical school [1970s], MMC was still the place you went to if you had complicated illnesses. Even private doctors would do hernias and gall bladders outside. But if you needed cancer surgery… they'd say: 'Hey, listen come to the government hospitals. They are better equipped to do all those. We are academic centres.' (Chennai doctor 2)

Another physician echoed these sentiments: '…I think more and more people joined medical college in Madras. It was considered to be, you know, the place to go to. So it had that reputation of being something, even right from the first days…' (Chennai doctor 7[20]).

---

[20] Interviewed on 10 May 2010.

Beyond a long-standing large medical infrastructure in and around Chennai in terms of medical education and large government hospitals, Chennai also became famous across the course of the twentieth century because of its large number of private nursing homes, run by prominent physicians. Chennai was notable for other large hospitals such as Southern Railway Headquarters Hospital in Perambur, Vijaya Hospital (established 1972), and others such as KJ Hospital and MV Diabetes Hospital.[21] In particular, noting private nursing homes such as Pandalai Nursing Home, Sundaravadam Nursing Home, and Kumaran Nursing Home, nearly every doctor whom I interviewed mentioned that Chennai's Poonamallee High Road came to be called 'India's "Harley Street"' among doctors and patients across India, particularly from the 1960s.

These hospitals themselves were part of a larger regional and national trend of an expanding private medical sector, beginning c. 1960. As Bhat observes, 'private healthcare expenditure in India has grown at 12.5% per annum since 1960'.[22] In rural India, the number of small private treatment facilities increased threefold between 1984 and 1992.[23] Similarly, in the small city of Mangalore in southwest India, the number of moderate-sized nursing homes jumped from 6 in 1986 to 20 in 1994 to 32 in 1998.[24] Further, as Nicher and Van Sickle explain:'In the 1980s, small private labs began springing up in towns and cities....'[25]

As home to these highly trained physicians in their private practices and at government institutions, Chennai came to be well known as excelling in particular specialist areas of medical expertise. One doctor summed up what many others noted:

[21] 'Is Madras Fast Becoming...The Medical Capital?' *Madras Musing* 1, no. 16 (1991).

[22] Ramesh Bhat, 'Regulating the Private Healthcare Sector: The Case of the Indian Consumer Protection Act', *Health Policy and Planning* 11, no. 3 (1996): 272.

[23] Ramesh Bhat, 'Characteristics of Private Medical Practice in India: A Provider Perspective', *Health Policy and Planning* 14, no. 1 (1999): 27.

[24] Mark Nichter and David Van Sickle, 'The Challenges of India's Health and Healthcare Transitions', in *India Briefing: Quickening the Pace of Change*, ed. Alyssa Ayers and Philip Oldenburg (Armonk, New York, and London: M E Sharpe, 2002), 184.

[25] Nichter and Sickle, 'The Challenges of India's Health', 185.

Historically, Chennai is the health care capital of India ... for whatever reason the primary centre is always started in and around Chennai.
Cardiac units, neurosurgical units, orthopaedic units; anything that starts
in India and healthcare first seems to be able to kick off in Chennai
and then to somewhere else. Dr B. Ramamurthy was the legend of
his time—a first world class neurosurgical centre that he put up in the
seventies.... Like that, the Cancer Institute in Adyar still has a reputation for being a good oncology centre.... In opthomology, [Sankara]
Nethralaya is a world class centre. (Chennai doctor 2)

Additionally, many physicians noted that medicine practiced in
Chennai had a reputation for a high level of professional ethics combined with relatively low costs. One doctor observed: 'A kind of good
temperament is there in most of the senior doctors in Chennai, they
want to be helpful not necessarily just make money' (Chennai doctor 3[26]). Another doctor spoke at length of the city's strong 'bed-side
manner' reputation as part and parcel of a broader set of regional cultural attributes:

I think the innate nature of the southern person is a lot more compas-
sionate and less money-minded. I mean it's a fact because, a person from
the north, by and large, we have come across so many people from there
who have come here who say that definitely you can get an appoint-
ment the very next day, or the following day here, despite however busy
the doctor is. And you actually get good medical care. And [Chennai
doctors] are not brusque and they don't push you off. So that kind of
thing still is there, I think, in the south, if you were to compare it with
the north. Because you had quality care at an affordable price here.
(Chennai doctor 7)

Along similar lines, another doctor elaborated on this theme in terms
of professional probity: 'Medicine really exploded during the post-war
years. And Madras had the reputation that it tended to be a little bit
more conservative. The Bombay person is always a little bit more of
an entrepreneur' (Chennai doctor 12[27]). Another doctor compared the
relative costs of medical treatment across the metros:

Whereas Maharashtra has always been the realm of private medicine...
It has always had the reputation of being phenomenally expensive.

[26] Interviewed on 5 July 2010.
[27] Interviewed on 11 July 2010.

Bombay doctors will charge three to four times what they will in Chennai. And somehow they did not have the overall professional competence that Chennai doctors did. (Chennai doctor 2)

1b. *Apollo came up at a time when government support for health care for ordinary Indians was flagging*

As asserted previously in 1a, this claim is difficult to sustain in the face of overwhelming evidence to the contrary. This is chiefly because there was a substantial government health infrastructure in Tamil Nadu before the establishment of either Apollo, or any other corporate hospital in the city.

As part of a larger project of widening and strengthening the health infrastructure in India, Primary Health Centres and subcentres were introduced as the 'Rural Health' part of the 'Minimum Needs Programme' in the Fifth Five-Year Plan (1974–8). As Duggal explains: 'During the 1980s, the public health spending peaked and this was reflected in major health infrastructure expansion in rural India via the Minimum Needs Programme'[28] (Duggal 2004, n.d.). Within this programme of health infrastructure building, Tamil Nadu was particularly successful. As Muraleedharan, Dash, and Gilson narrate: 'Tamil Nadu embraced the concept wholeheartedly and built the facilities much faster than almost all other states.'[29]

But there is also the possibility that Tamil Nadu was at least in part able to capitalize on the rural health agenda of the Minimum Needs Programme because the state was already undertaking a robust programme of health planning ahead of the implementation of the Minimum Needs Programme. K.S. Sanjivi (doyen of Chennai's voluntary health sector, b. 1903–d. 1994) claimed in 1973 that Tamil Nadu was one of the few states where all the primary health centres were actually provided with the medical and paramedical personnel needed.[30] Indeed, in 1973 Sanjivi explained:

[28] Ravi Duggal, 'Tracing Privatisation of Healthcare in India'.

[29] V.R. Muraleedharan, U. Dash, and L. Gilson, 'Tamil Nadu 1980s–2005: A Success Story in India', in *'Good Health at Low Cost' 25 Years On: What Makes an Effective Health System?* ed. D. Balabanova, M. McKee, and A. Mills (London: London School of Hygiene & Tropical Medicine, 2011).

[30] K.S. Sanjivi, 'Excellent Health and Medical Services', *Times of India*, Delhi, 22 May 1973.

The government of Tamil Nadu was the first to constitute a state planning commission with a task force on health ... presided over by Malcolm Adiseshiah.... [It] divided itself into working parties to consider in depth the problems of health services, medical education, family planning, nutrition, sanitation, the role of voluntary organisations and indigenous medicines including homeopathy. It handed over its report to the Chief Minister of Tamil Nadu, M Karunanidhi, in 1972.[31]

Taken together, the pre-existing government health infrastructure did much to contribute to the growth of private health care in Chennai—chronologically *preceding* the establishment of Apollo. Whereas Apollo may constitute a part of these larger trends, it would be anachronistic to claim that it could have served as a catalyst for them. In interviews, many other doctors also pointed out that this long-standing health infrastructure in and around Chennai in fact created client base for private medicine. Many doctors pointed out that, particularly by the 1980s, the region's population was sensitized to keep good health care 'habits', such as visiting biomedical doctors to address ill health concerns. Similarly, the long-standing relatively higher levels of development in Chennai and across Tamil Nadu meant that even before liberalization, Tamil Nadu was home to comparatively large middle class population with deep pockets and within striking range for seeking specialist care in Chennai.

1c. *Apollo triumphed because it provided what was up till then unavailable in India or for Indians*

This aspect of the Apollo myth claims that for all except the very, very rich, good quality health care was out of reach for Indians. The corollary claim is that those who could afford international travel went either to the UK or to the US for specialist, life-saving treatment. The following is an excerpt from an interview with a Chennai doctor:

> **Chennai doctor** 7: ... I mean, they were doing excellent work. But there was always this thing that if you were a VIP you got good treatment. Whereas if you were a common man, you didn't get such good treatment. And the hospital could have been a lot more clean. So that was one thing that did put people off. Now suddenly here was a place where everybody could go to. I mean not everybody—people who could afford it, and who did not want to go to a GH could come here.

---

[31] Sanjivi, 'Excellent Health and Medical Services'.

SH: Instead of going abroad.

CD7: Yes. Now you talk about the heart. Everybody who needed a bypass would go to the US. Now suddenly here was a place that one could go to. You didn't have to go there.

However, what the claim that Apollo triumphed because it provided what was up until then available for ordinary Indians generally fails to take into account is the additional circumstance that in the mid-1980s, for Indians, the cost of international travel rose astronomically. This was because of changes in the exchange rate, and, in particular, radical devaluing of the rupee, particularly by the late 1980s. It was at precisely the same time that overseas medical travel became prohibitively expensive that Apollo began to announce dividends.

2. The emergence and rise of Apollo provided a new model of health care delivery in India

This claim is predicated on the corollary claim that Apollo provided a new model of health care delivery in India because Apollo was quickly emulated by many others in private health care sector. Certainly, the story of Apollo can be told as the story of the emergence and rise of *one* corporate hospital in India. Apollo's promoters, from the beginning, planned Apollo not as a singular institution, but as a *chain* of hospitals.[32] What is less clear, however, is the degree to which the appearance of Apollo in fact served as a *catalyst* for the successful emergence of other large private hospitals across India.

One of the claims this is based on is that Apollo was an immediate success. However, not only did Apollo take at least five years before it made any dividends, there was a substantial lag-time between the inauguration of Apollo in 1983 and the opening of other large Indian hospital chains that continue till today. Nevertheless, it is the case that as early as 1991, Chennai was notable as a 'corporate health care' city for India. Yet this was the case because the city had a grand total of four large private hospitals. As one journalist celebrated: 'Madras is the new "mecca of medicine"…In the last five years the hospital services sector has boomed in this city, though "for profit" hospitals exist

---

[32] *Times of India*, 'Madras Hospital to Be Run as Public Limited Company'.

elsewhere in the country, Madras is the only city with four corporate hospitals'.[33]

Indeed by 1995, Chennai had six corporates: Tamilnad Hospital, Devaki Hospital, Malar Hospital, Dr Agarwal's Eye Hospital (which went public in 1994), and Chennai Kaliappa Hospital, in addition to Apollo. However, the business model for starting corporate appears to need very deep pockets. Many doctors observed that 'the gestation period for a hospital is five to seven years, minimum, before it can make a profit' (Chennai doctor 6[34]). As one physician explained:

> When you borrow money [for a hospital], you're asked to repay like an industry in like five years. [But] you cannot pay back in healthcare in the five years. Absolutely impossible. So what then happens is that people take the massive amount of money. [But] modern medical technology depreciates in four years. At the end of the fifth year, you have junk, it's scrap.... (Chennai doctor 2)

The following tale of Tamilnad Hospital illustrates how, while, it was one thing to open a corporate hospital, it was quite something else to keep it going or even to turn a profit.

From Tamilnad to Global

Tamilnad Hospital was incorporated in 1984 by a US-based non-resident Indian, Dr C.P. Velusamy. In 1985, Tamilnad Hospital became a public limited company. In 1991, Tamilnad Hospital issued public shares in order to finance the cost of setting up what was at that time described as 'the first phase'—a 250-bed hospital in Perumbakkam, for the time quite remote south in suburban Chennai.[35] However, after a protracted labour dispute, in 2000, Tamilnad Hospital faced a mass walk-out of physicians.[36] After this labour unrest and after a lengthy

---

[33] Gouri Umashankar, 'The Marketing of Health', *Times of India*, Delhi, 15 January 1995.

[34] Interviewed on 11 May 2010.

[35] 'Tamilnad Hospital Ltd, Company Profile', *Indiainfoline.com*, no date, http://www.indiainfoline.com/Markets/Company/Background/Company-Profile/Tamilnad-Hospital-Ltd/523443 (accessed 24 September 2012).

[36] S. Gopikrishna Warrier, 'Tamilnad Hospital Impasse May End: Management Change Likely', *Hindu Business Line* (Internet edition), 6 July 2000, http://www.hindu.com/businessline/2000/07/07/stories/14076012.htm.

delay to plans to start a medical college jointly with the nearby Kanchi *math*, Tamilnad Hospital folded. By 2003, the hospital was taken over by the Kanchi *math* through its deemed university at Kancheepuram. The hospital was rechristened 'Sri Kanchi Kamakoti Sankara Medical Hospital'.[37] Yet, in 2006, Sankara Hospital admitted defeat in being able to turn a profit in the venture and applied to sell the hospital, explaining that in its expansion to 450 beds on the 46 acre site, it had become untenable financially. In 2007 in an all-cash deal worth INR 257 crore, Sankara Hospital was bought by Global Hospitals and became Global Health City, which it remains today.[38]

3. At its heart, Apollo is a patriotic project

This aspect of the Apollo myth is one that is promoted the most heavily by the Apollo Hospitals Group itself in its publicity materials. As Reddy regularly emphasizes in the many media interviews he gives, '…bringing the best healthcare within the reach of every patient is our mission and [at Apollo] we are determined to make it a reality'.[39] What this claim addresses is an implicit criticism of for-profit medicine as profit-driven, and therefore morally suspect in a nation wracked by long-standing enduring poverty. In framing the business as national service, this claim neutralizes this criticism. Indeed, at points, the Indian entrepreneur's very pursuit of profit (whether in health care or other ventures) is presented as patriotic in and of itself—particularly in terms of the service it is seen to provide to middle-class consumers. As one of Reddy's recent interviewers wrote: '[Reddy's] is the story of one man who set out to revolutionise the unaddressed health-care needs of a section of India's growing middle

---

[37] V. Balasubramanian, 'Faith Healing: Kanchi Math to Take Over Ailing Hospital', *Economic Times*, Delhi, 14 April 2003, http://articles.economictimes. indiatimes.com/2003-04-14/news/27553032_1_tn-hospital-general-hospital-corporate-hospital (accessed 24 September 2012).

[38] Sanath Shetty, 'Global Hospitals Buys Chennai-based Sankara Hospital for Rs 257 crores'. *Dealcurry.com*, Mumbai, 5 April 2007, http://www. dealcurry.com/20070405-Global-Hospitals-buys-Chennai-based-Sankara-Hospital-for-Rs-257-crores.htm (accessed 19 February 2013).

[39] Prathap Reddy interview by Ranjini Manian, 'Heart to Heart', *Culturama* 1, no. 5 (July 2010): 16.

class. It is a tale of manoeuvring through difficult bureaucratic and complex medical systems.'[40]

Much of the retrospective publicity concerning Reddy and opening the first Apollo rehearses claims about a specific set of prime ministerial meetings with, first, Indira Gandhi, and, later, Rajiv Gandhi. These always include a version of the following story: 'I told Mrs Gandhi only the rich and powerful get access to health care and she really gave the first impetus by telling everybody that "Here's a man who wants to reverse the brain drain".'[41] And indeed Indira Gandhi, who only lived to see the first Apollo open less than a year, is regularly credited with evaluating the effect of the first Apollo. Of this, another of Reddy's interviewers wrote:

> The new hospital attracted the best medical talent, including eminent non-resident Indian doctors who returned to India from hospitals in the US and UK. This prompted then Prime Minister Indira Gandhi to remark, 'Dr Reddy you have brought talent back to India and reversed the brain drain'.[42]

But Reddy regularly remarks that, additionally, '...the man who really changed the face of healthcare in this country with his vision and clarity was none other than Rajiv Gandhi—by opening up hospitals to funding and other opportunities'.[43]

However in 1982, when Reddy was preparing to open Apollo, his claims for the hospital's national role were substantially different from this subsequent story. At the time, one newspaper reported:

> A hospital being built under the corporate sector here expects a steady flow of rich Arab clients and a huge inflow of foreign exchange, since the Arabs are not satisfied with the facilities offered in the Bombay hospitals. Dr Prathap Reddy, chairman of the company behind the venture, told newsmen here yesterday that many rich Arabs had told him that they wanted to be picked up from the airport to the hospital and all investigations and treatment should be done under one roof, regardless of cost.[44]

---

[40] Gupte, 'Transforming India's Health-care Landscape'.

[41] 'Interview with Prathap Reddy, Apollo Hospitals', *Indian Medicos*.

[42] 'Interview with Prathap Reddy, Apollo Hospitals', *Indian Medicos*.

[43] 'Interview with Prathap Reddy, Apollo Hospitals', *Indian Medicos*.

[44] *Times of India*, 'Hospital to Earn Foreign Exchange', *Times of India*, Delhi, 1 December 1982, 6.

Such pronouncements chime with Reddy's slightly earlier announcement that the government recognized Apollo's venture as a 'core
economic activity' because government grasped the potential of health
care to net foreign exchange.[45] Nevertheless, the point is that this quote
is at odds with the vision of Reddy and Apollo that has been com-
memorated subsequently. Instead, Reddy received a Padma Bhushan
in 1991 (India's third highest civilian honour), a Padma Vibhushan in
2010 (India's second highest civilian honour) and the Indian Postal
Service launched a commemorative Apollo Hospitals stamp in 2009.

4. **In opening Apollo, Reddy changed state practices single-
handedly**

This claim is based on assertions that Reddy effected these changes
by dint of his personal charisma which he used to finesse the Delhi
Durbar under successive 1980s Prime Ministers and by the sheer per-
suasiveness of his argument that his was a national/populist project.

In these accounts, Indira Gandhi and Rajiv Gandhi also figure prominently: 'Banks were not willing to fund hospitals. Apollo
approached the Centre and found a patient listener in the then prime
minister Indira Gandhi. Health care sector gained industry status and
access to financial markets.'[46] Another interviewer claimed that when
Rajiv Gandhi was Prime Minister in 1989,

> ... on Reddy's representation, the former amended in three days in the
> Parliament and removed all hardships to liberal funding. And so the cost-
> liest equipments made inroads into Indian hospitals and were equipped
> on par with the Western. Rajiv Gandhi also made a tax exemption of
> Rs 10,000 [on medical equipment].[47]

Finally, another interviewer risks over-egging even Reddy's and
Apollo's own claims: 'Often referred to as the father of modern

[45] *Times of India*, 15 April 1982, p. 6.

[46] 'The Financial Chronicle Highlights Apollo Hospitals', *Apollo Hospitals
Enterprise Limited*.

[47] Gopal Ethiraj, 'Sunday Celebrity: Dr Prathap C Reddy who
Revolutionised Health Care System in India', *asiantribune.com*, 2 January
2010, http://www.asiantribune.com/news/2010/02/01/sunday-celebrity-
dr-Prathap-reddy-who-revolutionized-health-care-system-india (accessed 16
February 2013).

healthcare in India—after all, he revolutionised healthcare in India when the country was mired in babudom. Doctor-entrepreneur Prathap C. Reddy remains unfazed by such adulation.'[48] Of the first Apollo, Reddy himself is quoted as explaining that

> ... securing licenses to import 370-odd medical equipment for the hospital itself took two years. Meanwhile, lowering of import duty on life-saving medical equipment also helped private healthcare during the pre-reform era. The duties came down from 100 per cent to 5–6 per cent over the years.[49]

Yet these claims ignore and obscure the fact that there was a pre-existing economic climate, already changing in the ways that Reddy takes credit for. In particular, the growth toward liberalization seen under, first, Indira Gandhi in the early 1980s and followed up by Rajiv Gandhi in the late 1980s. This aspect of the myth also obscures the increasingly activist role of associations such as the Federation of Indian Chambers of Commerce and Industry (FICCI) and the Confederation of Indian Industry (CII).

5. Apollo was an immediate success

In popular retrospection (and in the publicity circulated by Apollo Hospitals), in terms of therapeutic outcomes and profit margins, the success of Apollo is presented as an *immediate* success. As one doctor emphasized in an interview: '[Apollo] was a place where you could be confident you would get every kind of treatment under one roof. And it was available for a price, but it was there. The quality was there. That was right from the beginning. It was a foregone thing. It just took off' (Chennai doctor 7). Much of this attributes success to the Reddy's sheer visionary nature. As another doctor declared: 'Apollo succeeded because Reddy could see what was coming' (Chennai doctor 8[50]). But as is clear from both Apollo's own attempts to secure funding through further public share issues to underwrite further expansion, alongside the struggles of other hospitals to thrive within the same Chennai market, this was not the case.

---

[48] 'Interview with Prathap Reddy, Apollo Hospitals', *Indian Medicos*.

[49] 'Financial Chronicle Highlights Apollo Hospitals', *Apollo Hospitals Enterprise Limited*.

[50] Interviewed on 11 May 2010.

MEDICAL TECHNOLOGIES AND THE
CORPORATE HOSPITAL

Finally, many doctors claimed that part of the significance of Apollo was that, 'Apollo became synonymous with high-tech medical care' (Chennai doctor 6). Although it is not something *unique* to Apollo, many doctors whom I interviewed spoke at length about the relationship between technological innovation in diagnostics and clinical treatment and claimed that the emergence and uptake of medical technology has largely been responsible for the commodification of health care in India. To rehearse this argument briefly, for India, beginning in the 1980s, capital intensive medical technologies (such as MRI scanners, CT Scans) came to dominate clinical diagnostics and hospital care. Practitioners who purchased these machines were under pressure to make their money back as quickly as possible. This was because this equipment required substantial outlay, and purchases were often financed through debt. Whereas under the earlier model of care in smaller private nursing homes and pathology labs, a regime of 'interpretation charges' came to dominate the relationships that connected practitioners' diagnostic practices to the prospect of broader financial gain. Under the large multi-speciality private hospital model, hospitals took on more debt, but consolidated clinical diagnosis and care under one roof. Having taken on debt, hospitals needed to net more revenue. This in turn drove more investigations with specialist equipment, which in turn led to more diagnoses of care and treatment, and often, further investigations. In other words, simply purchasing medical equipment was not enough, use (and what is suggested here is deliberate overuse) of one piece of equipment not only had to pay for itself but the revenues from one machine were designed to be put toward the purchase of the next one. This has been particularly the case in a landscape of rapidly changing technologies of care that quickly rendered medical equipment out-of-date if not completely obsolete.

While this analysis may go far to explain the commodification of medical technologies within health care, it is certainly also the case that not only the use, but the very purchase and installation of new pieces of high tech medical equipment were used by hospitals to generate revenue. As a result, from the 1980s onwards, newspapers are filled with news stories that are little more than announcements of equipment purchases (and later hospitals' paid advertisements announcing VIP inaugurations of said pieces of equipment). So the very existence of the equipment itself is used to generate revenue, professional referral, and patient footfall. [51]

Newspaper articles announcing the plans for the hospital and the financing arrangements also invariably included a list of the new medical technologies and machines that would be available at Apollo. Sometimes, entire (if brief) newspaper articles were devoted to such machines. For example: 'The latest generation whole-body scanner, the SOMATOM DR-3 from West Germany, has been delivered to the Apollo Hospitals here.... Medical professionals all over the world are reported to have welcomed it as one of the most powerful diagnostic tools available.'[52] Similarly, another article reported:

> A non-surgical alternative to bypass surgery for patients with coronary artery disease—coronary angioplasty—is now available in the city. The non-surgical treatment is available at Apollo hospital. The hospital authorities say only stray cases of angioplasty have been done in a few places in the country, though the technique has caught up all over the US and Europe.[53]

In this way, simply the existence of the equipment itself is seen to be a marketing device of the 'latest, greatest' techniques in medical diagnostics and treatment.

MEDICAL MYTHOLOGIES AND POST–LIBERALIZATION FUTURES

Alongside biotechnology and information technology, corporate health care is given pride of place within India's current 'sunshine story'. These industries are taken to be examples of India's capacity to deliver 'science'. They are given much credit in accounting for the nation's recent economic successes. Indeed, according to a recent

---

[51] One of the 'laws' of economics that holds is Say's Law: that supply creates its own demand. This is particularly the case in medical care due to its unique feature, namely asymmetry of information. Which patient would refuse a CT scan if her doctor tells her it is necessary?

[52] *Times of India*, 'Whole Body Scanner for Hospital Chain', 15 May 1983.

[53] *Times of India*, 'Angioplasty Now Available in Madras', 4 May 1986.

KPMG report, rising income levels, changing demographics, and shifts in disease profile are expected to double the size of health care spending by 2012.[54]

The industrial barons leading these fields both drive new economic policies in a broader privatization of science, and profit from these new policies. Returning to the major new initiative that was referred to at the beginning of this chapter, the proposals for Universal Health Care in India, one final point about the form and meaning of medicine and myth under liberalization is in need of critical scrutiny. That is, the claim that the corporate 'multi-speciality' hospital both served as an innovation in health care delivery, and the corollary claim that the corporate 'multi-speciality' hospital provides an extremely broad range of high-quality medical services and, as such, a template for health care delivery to the nation.

Particualrly in interviews with doctors, I encountered widespread agreement with the view that multi-speciality corporate hospitals provided innovation in health care delivery. For example, one doctor explained the significance of multi-speciality large private hospitals:

Suppose a specialist—I am talking about 20 years back—suppose you are an eye specialist. You will have eye hospital. Or, you are a surgeon. You will have a surgical hospital. But the corporates changed that. Apollo hospital changed that concept. They said, 'All departments under one roof'—that was the concept. (Chennai doctor 10[55])

Another doctor echoed this view:

[Apollo] was a place where you could be confident you could get every kind of treatment under one roof...Suddenly people found that here was a place that, you know, they were having all *kinds* of specialities. That was the first hospital that actually even had specialists coming in. (Chennai doctor 7)

Listening to these accounts, I failed to see what was so innovative about this. Surely, I thought to myself, the concept that one hospital could treat an entire corpus of ailments, was the *foundational idea*

---

[54] KPMG, 'Emerging Trends in Healthcare: A Journey from Bench to Bedside', 17 February 2011, http://indiainbusiness.nic.in/newdesign/upload/news/Emrging_trends_in_healthcare.pdf.

[55] Interviewed on 8 March 2010.

of hospital medicine, as it emerged in the late eighteenth and early nineteenth centuries. Many claim that the multi-speciality hospital provided something new, but surely this was simply a shinier imitation of already-existing government and charitable institutions, themselves also based on a long-standing model of comprehensive clinical investigation and treatment.

But the problem with this misconception is beyond a basic amnesia for medical history. The problem with this celebration of the corporate multi-speciality version of hospital care is that, over the past three decades in India, it has succeeded only selectively. Indeed even among doctors who praised the multi-speciality corporate hospital's supposed 'innovative' model of care, there was the simultaneous recognition that most multi-speciality hospitals succeeded both therapeutically as well as financially because they made very selective and strategic choices about their investments in specializations. These specializations allowed for very high success rates for very specific procedures that could also deliver a high patient through-put and a corollary income stream. Take the example of the Southern Railways Headquarters Hospital in Chennai (henceforth 'Railways'). Railways offered excellence in heart-related ailments. However, because of the wide cross-section of the population that this hospital was built to serve, heart specialists there were able to develop expertise not only in coronary by-pass surgery, for example, but also in the much riskier areas of paediatric cardiac surgery, or in heart ailments suffered disproportionately by the poor (for example, rheumatic heart diseases that do not necessitate open heart surgery).

The significance of mistaking the comprehensive care that the corporates claim to deliver for a more genuinely comprehensive care, is when one considers what the corporates are being asked, and are poising themselves to deliver for the general population of India under Universal Health Care proposals. Neither the corporates nor the Government of India suggest that corporates should become involved in public (or 'preventative') health care. As one profile of Reddy explained: 'As Dr Reddy himself acknowledges, primary healthcare should be the responsibility of the government, which has both the resources and manpower to reach all parts of the country.'[56] This is echoed by India's political leaders today. As one chief minister put it,

[56] Gupte, 'Transforming India's Health-care Landscape'.

'To provide a reliable healthcare to one and all is, no doubt, a big challenge. It is imperative to work together as only governmental efforts will not suffice.'[57] Indeed Reddy himself claims: 'My vision for the Apollo Hospitals Group is to touch a billion lives, and I am sure we will fulfil the dream.' It is hard to see that he wants to touch all parts of these lives' bodies,[58] just the revenue-generating bits.

[57] KPMG, 'Emerging Trends in Healthcare', 1.
[58] Manian, 'Heart to Heart', 16.

Part 3

National Techno-science and
Promising Bodies

7 The Globalization of Reproduction in India

From Population Control to Surrogacy

Mohan Rao★

Could men but come awake—enchantments keep
Their noblest faculties held fast in sleep
And frightful dreams and real fears, alas!
Before their soggy haunted vision pass
Not least the reverend Thomas Malthus with his trick
Of killing conscience with arithmetic.

—James MacAullay[1]

It has long been common sense among government officials that India has a population problem: too many people. This common sense was created and distributed more than 200 years ago, encapsulated in the thesis of English economist and clergyman Thomas Malthus. What it did do successfully was to exonerate the various means by which industrial capital in England was pauperizing peasants by taking away their lands and livelihoods, thereby creating a population 'surplus' to capital's requirements. In the nineteenth century, such economistic population thinking became meshed and intertwined with eugenic

★ I am deeply grateful to Sarah Sexton for breathing shape to this chapter.

[1] James MacAullay, 'A Vision of Ceremony' (undated), quoted in Colin Clark, *Population Growth and Land Use* (London: Macmillan, 1968), 157.

and racial theories: the population problem was one of too many people of the wrong sorts.

In the post-independence period, the main 'solution' to the problem so constructed has been contraception and sterilization primarily targeted at disadvantaged women in clinics throughout the country where family planning overrides the provision of other health care services while the determinants of good health—food, water, shelter—are neglected.

Today, public health care provision has declined still further since India embarked on its neo-liberal economic policies in the 1990s, providing tax breaks for high-tech for-profit health care services at the same time. But in a twenty-first century global twist of overpopulation, eugenic, and neoliberal logics combined, high-tech, profit-motivated medicine has poorer Indian women in its maws once again, not to stop them from giving birth but to encourage them to do so via the merged in-vitro fertilization (IVF), assisted reproductive technology, and surrogacy industries. These women now turn waste into gold! The babies to whom they give birth, however, are commissioned by and handed over to couples, often from abroad, who have provided the sperm, and perhaps the egg as well, and who have obtained enough money to pay the assisted reproductive technology (ART) and surrogacy clinics for their services which are more expensive or banned in their own countries.

Outside of this international high tech industry, health and health care services dwindle. The picture is bleak. 56,000 women die each year in India during childbirth. 309,000 babies die each year on the day they are born; 876,200 babies die in the first month of birth, while 165,400 do not live to see their fifth birthdays. Rather than addressing these, the surrogacy business is invariably justified by claims that the women who carry and bear children for others at least get some money out of it, in what is called a win–win situation.

This chapter shows how India's thriving surrogacy industry that aims to give life to a few particular people is premised on at least two centuries' influential theories and powerful practices—population, eugenics, and neo-liberalism—that take life and livelihoods away from the majority of others.

This chapter first describes how nineteenth century overpopulation theories dovetailed with India's surplus being sucked into Britain's empire.

It goes on to show how nineteenth and twentieth century eugenic arguments became intertwined with those of the idea of a surplus population and the resulting family planning goals that skewed health care services. It concludes by setting the emergence of India's high-tech medicine in the context of its pursuit of neo-liberal economics over the past three decades. All three ideas and practices mesh in what has become a global surrogacy industry with India as its major hub.

OVERPOPULATION

India was first integrated into the world of capitalism after the Battle of Plassey in 1757. The colonial loot of the 'jewel' in Britain's crown was both immediate and staggering. The treasure taken out of India between Plassey and Waterloo, that is, between 1757 and 1815, has been estimated at an astounding £500 million to 1,000 million.[2] Its impact on the industrial revolution in England was just as immediate as noted, 'The Bengal plunder began to arrive in London, and the effect appears to have been instantaneous.... At once in 1759, the Bank (of England) issued 10 and 15 pound notes ... and the industrial revolution began with the year 1760'.[3]

Thus with India's provision of a primitive accumulation of capital for England's industrialization began the process of her own impoverishment. The country was 'systematically deindustrialised'[4] as it became a market for Lancashire cottons: in 1820 India took 11 million yards; by 1840 its imports had jumped more than ten-fold to 145 million yards. As spinning and weaving in urban India became redundant, the artisans became destitute; the immiserized rural artisans were ready victims for the periodic famines that loomed over India in the nineteenth century, their plight exacerbated as India's wheat became London's bread basket.

Indeed, British agricultural policy—of commercialization and revenue extraction through the Permanent Settlement Act—impoverished

---

[2] Simon Digby, quoted in Prabhat Patnaik, 'On the Political Economy of Underdevelopment', *Economic and Political Weekly* 7, nos 4–6 (1973): 197–212.

[3] Brooks Adams, quoted in Patnaik, 'On the Political Economy of Underdevelopment', 197–212.

[4] Eric J. Hobsbawm, *The Age of Revolution* (London: Abacus, 1977).

and dispossessed vast sections of the Indian peasantry[5] who could not go to work in mills, industries, and factories, as their English counterparts did because British free trade policies actively hindered industrialization in India. The sub-continent was thus integrated into the world economy as an exporter of primary commodities with a massive, virtually stagnant agricultural sector and no industrial sector to speak of.

The French catholic priest, Abbe Dubois, in India to escape the terror of the French Revolution,[6] surveyed the destruction of the Indian weaving industry and the consequent pauperization of her artisans, linking the process to the dark satanic mills of England. He nevertheless remarked that the 'chief cause of their misery'

> ...is the rapid increase of population. Judging by my own personal knowledge ... of Mysore and the districts of Baramahl and Coimbatore, I should say that they increased by 25 per cent in the last 25 years.... Some modern political economists have held that a progressive increase in the population is one of the most unequivocal signs of a country's prosperity and wealth. In Europe this argument may be logical enough, but I do not think that it can be applied to India; in fact, I am persuaded that as the population increases, so in proportion do want and misery. For this theory of the economists to hold good in all respects the resources and industries of the inhabitants ought to develop rapidly; but in a country where the inhabitants are notoriously apathetic and indolent, where customs and institutions are so many insurmountable barriers against a better order of things, and where it is more or less a sacred duty to let things as they are, I have every reason to believe that a considerable increase in the population should be looked upon as a calamity rather than as a blessing.[7]

This idea that poverty in India was because the country had too many people quickly became 'common sense' for colonial administrators at Haileybury College, the East India Company's training ground

---

[5] Patnaik, 'On the Political Economy of Underdevelopment', 197–212.

[6] Abbe Dubois travelled extensively through South India in the aftermath of the 1789 French Revolution, when many clerics were killed. His journals were subsequently purchased by the East India Company to understand better the people they were trying to rule.

[7] Jean-Antoine Dubois, *Hindu Manners, Customs and Ceremonies*, trans. H.K. Beauchamp (Oxford: Clarendon Press, 1906), 93–4.

for its employees before they set sail to India. Indeed, Malthus himself was appointed Professor of Political Economy at this college, a position he held for some 34 years.

One Haileybury trained administrator was Richard Strachey, who headed the Famine Commission set up to survey the devastating famines of 1877. He appears not to have been a heartless man; he argued, for instance, that it was the duty of the government to provide famine relief, a fiscal responsibility that he did not believe was overwhelming. After all, famines diminished the revenues derived from taxation. But the lack of industrialization and alternative sources of employment needed to be addressed as a way of tackling a population surplus to requirements. 'With a population so dense as that of India ... the numbers who have no other employment than agriculture, are in large parts of the country greatly in excess of what is really required for the thorough cultivation of the land'.[8]

Nevertheless, echoing Malthus and his training, he argued that relief must not encourage people to become dependent on such welfare, nor their further breeding, but, should instill thrift in them. Relief must be designed so as to reach the deserving poor:

> All such advantages tend ... to favour the increase of the population, and to add to the pressure on the means of subsistence. It becomes, therefore, especially important that the Government, when it has to deal with calamities such as famines, should so frame its measures as to avoid every tendency to relax in the people the sense of the obligation which rests on them to provide for their own support by their own labour, to cultivate habits of thrift and forethought, and as far as possible to employ the surplus of years of plenty to meet the wants of years of scarcity. The great object of saving life and giving protection from extreme suffering may not only be as well secured, but in fact will be far better secured, but in fact care be taken to prevent the abuse and demoralization which all experience shows to be the consequence of ill-directed and excessive distribution of charitable relief.[9]

Strachey considered it given that relief cannot reach the undeserving poor, who starve by their own failings:

---

[8] Government of India, *India Famine Commission* (Chairman: R. Strachey, George Edward Eyre and William Spotterwoode, London, 1880), 34.
[9] Government of India, *India Famine Commission*, 35.

There must always be suffering and want which will escape notice; and however extensive be the measures of public aid, and however reasonable be the terms on which it is offered, there will always be classes who, from fixed habits or social institutions of various sorts, or from their personal character or ignorance, will neither help themselves nor be helped, and who, though they suffer from extreme want, will linger on without applying for or accepting relief till it is too late to save their lives.[10]

Moreover, relief, or any other government policy, must not interfere with free trade other than to facilitate it. The government's goal, Strachey believed, should be:

To maintain a policy of non-interference with the ordinary operations of trade unless in some very exceptional condition of affairs when there may be evidence that without such interference the supply of food will not be maintained; but to keep a constant watch over the food supply of the people in time of threatened or actual scarcity, and to remove any impediments in the way of the free movement of trade.[11]

The conclusions of the Famine Commission were inevitable, given their logic: 'The statistical returns made certain what has long been suspected, that starvation and distress greatly check the fecundity of the population.'[12]

That there were too many (fertile) people in India was a refrain a decade later in testimonies to another enquiry set up under Lord Dufferin to examine the extent and causes of poverty.[13] The Deputy Commissioner of Shahpur, Mr T. Wilson describes the causes of poverty as want of thrift and too rapid an increase of population, while the Secretary to the government of Bengal, one Mr P. Nolan, noted that poverty occurred due to high rates of rent too but that population growth had a decisive influence:

Simply because labourers and competitors for land are more numerous ... as it appears that the pressure of population on natural resources has a decisive influence.... In addition to the rapid increase of population,

---

[10] Government of India, *India Famine Commission*, 37.

[11] Government of India, *India Famine Commission*, 37.

[12] Government of India, *India Famine Commission*, 28

[13] Government of India, *Report on the Condition of the Lower Classes of Population in Bengal* (Calcutta: Revenue and Agricultural Department, 1888).

the high rates of rent in force in Behar, and the arbitrary methods of enhancement, have been quoted as causes of the poverty of ryots.[14]

Most commentators attributed poverty and deprivation to customs and ceremonies, 'indolence and spiritlessness', a 'mental paralysis' due to the climate that is 'paralyzing the vigour and enterprise of the race'. The exception to these sentiments came from Babu Bissessar Banerjee, Settlement Officer of Bogri and Kesiari, who noted, 'The importation of European piece-goods has impoverished the weavers. Many of them have given up their profession and now keep shops. Some are employed as servants. Most of them cannot afford to have two meals a day.'[15]

Analysis of the first all-India 1891 census observed, 'The conclusions arrived at by Malthus … in the main … have not been disproved.' But doubts are expressed too, as to whether the number of people are too many such that they will starve:

> We have every reason … to assume that the present rate of increase amongst the people of India is well within their means of subsistence. If maintained, which of course it will not be, it would be 75 years before the population doubled itself, and the problem of their support would then, no doubt, be a hard one for our successors.[16]

By 1911, however, these doubts had vanished. Famines and epidemics could actually do good, the government believed, ridding the population of her 'weaker' elements:

> The high mortality at the two extremes of life and among the weaker members of society left a population purged of its weaker elements and with constitution improved both physically and morally by the trials it had gone through. Though the population was almost decimated, though at one period nearly a fourth of the total population came on relief lists, though land went out of cultivation, cattle died, cheap crops took the place of valuable ones, while prices rose to levels never before attained, yet amidst all this hopeless depression and seemingly complete demoralization there emerged almost as if by a miracle a new spirit of vigour and energy. It had apparently needed a severe trial and

---

[14] Government of India, *Report on the Condition of the Lower Classes*, 29

[15] Government of India, *Report on the Condition of the Lower Classes*, 68

[16] Government of India, *Census of India, 1891* (General Report by J.A.Baines, Eyre, and Spotterwoode, London, 1893), 102.

tribulation to bring out qualities and energies which had so long lain intent during the anterior period of early existence.[17]

Between 1911 and 1921, the total population of India went down in part the consequence of the 1918–19 influenza pandemic. The commissioner of the 1921 Census, J.T. Marten, was rather sanguine about population growth, arguing that India's demographics continued to be dominated by, 'Nature's ordinary programme for the restriction of excess population'.[18] The 1931 Census, however, was the first to provide evidence of 'the grave increase in the population of this country'.

> The actual figure of the increase alone is little under thirty-four million, a figure approaching equality with that of the total population of France or Italy and appreciably greater than that of such important European powers as Poland and Spain. The population now even exceeds the latest estimate of the population of China, so that India now heads the list of all the countries in the world in the number of her inhabitants. This increase, however; is from most points of view a cause for alarm rather than for satisfaction.[19]

This report, at a time when the civil disobedience movement was growing, provided a number of commentators with the justification for continuing colonial rule.[20] Even though technical and medical advances were bringing down death rates, the director general of the IMS, John Megaw, observed, 'The people multiply like rabbits and die like flies: until they can be induced to restrict their rate of reproduction, there is no hope of doing much good by medical relief and sanitation, as the population is very nearly up to possible limit.'[21]

---

[17] Government of India, *Census of India 1911*, vol. 10: Central Provinces & Berar (Calcutta: Government Printing, 1911), 28.

[18] Government of India, *Census of India 1921, Part I: Report* (Superintendent of Government Printing, Calcutta, 1924), 47.

[19] Government of India, *Census of India 1931*, vol. 1 (Government Printing, Calcutta, 1933), 29.

[20] David Arnold, 'Official Attitudes to Population, Birth Control and Reproductive Health in India, 1921–1946', in *Reproductive Health in India: History, Politics, Controversies*, ed. Sarah Hodges (Hyderabad: Orient Longman, 2005).

[21] John Megaw, cited in Arnold, 'Official Attitudes to Population, Birth Control', 30

The focus on population growth as an explanation for Indian poverty held great appeal not only to the British colonialists, but also to certain sections of Indian society as well. For example, P.K. Wattal's influential work entitled *The Population Problem in India: A Census Study*, published in 1916, commences with Malthus' law of population and concludes that the 'alarming' growth of Indian population was responsible for widespread poverty and ill-health.[22] By the 1930s, 'significant layers of the native elites adopted (had) neo-Malthusian views'.[23] A promiscuous network of ideas, individuals, and institutions had created this juggernaut; no one, but no one, was untouched. This widespread 'common sense' became the bedrock for ideas of post-independence development, shaping the Left almost as much as the Right.

EUGENICS

As many of the quotes cited previously illustrate, overlaid and intertwined with these economic notions of overpopulation—too many people for the food available—were eugenic overtones that qualified the too many: those who were lazy, indolent, less intelligent, and so on. Eugenic ideas were articulated soon after those of Malthus but spread around the world in tandem with them.

Charles Darwin's *The Descent of Man*, published in 1871, provided a significant measure of inspiration to the birth of the Eugenics Movement.[24] But ideas of competitive struggle, natural selection, and the survival of the fittest, had frightening consequences when applied to the social arrangements of human societies not to mention deep ethical and moral implications. Racial purity and the improvement of the racial stock were the prime concerns of the Eugenics Movement, named as such by one of its founders, Francis Galton, a cousin of Darwin's, who pioneered the use of statistics on human populations. A.R. Wallace, who co-discovered the process of evolution with Darwin, argued that

[22] P.K. Wattal, *The Population Problem in India: A Census Study* (Bennet Coleman, Bombay, 1934).

[23] Bonnie Mass, 'An Historical Sketch of the American Population Control Movement', *International Journal of Health Services* 4, no. 4 (1974): 651.

[24] Germaine Greer, *Sex and Destiny: The Politics of Human Fertility* (London: Secker and Warburg, 1984).

it does not seem possible for natural selection to act (on Man) in any way so as to secure the permanent advancement of morality and intelligence for it is indisputably the mediocre, if not the low, both as regards morality and intelligence who succeed best in life and multiply fastest.[25]

He inspired Jane Hume Clapperton who published *Scientific Meliorism* in 1885.

The guardians of social life in the present day dare not be careless of the happiness of coming generations, therefore the criminal is forcibly restrained from perpetuating his vicious breed. The type will disappear whilst evenly balanced natures, the gentle, the noble, the intellectual, will become parents of future generations, and the purified blood and unmixed good in the veins of the British will enable the race to rise above its present level of natural morality. To promote the contentment of congenital criminals within their prison home, where they are detained for life, an alternative to celibacy might be offered, viz, a surgical operation rendering the male sex incapable of reproduction.[26]

Galton's passion was the collection of biostatistics and of data on the lineage of the pedigreed. He was firmly committed to the idea that only the brightest and best should be encouraged to have children. Eugenics therefore had two types of action on its agenda: the positive

[25] Cited in Greer, *Sex and Destiny*, 257

[26] Compulsory sterilization of a range of people considered unfit to be parents is of immense appeal to eugenists seeking a Utopia of the pure and meritorious. Tara Ali Baig, head of the Indian delegation to the 1974 UN Population Conference at Bucharest and the Chairperson of the Indian Council of Child Welfare, said,

Sterilization of one partner has to be made imperative where a man or woman suffers from hereditary insanity, feeblemindedness or congenital venereal disease: they must be barred by law from procreating children. This should have been done decades ago. If children's lives and future are to be protected, compulsory sterilization is necessary for many reasons.... After all considering the crime against children committed by irresponsible parenthood, compulsory sterilization is hardly punitive. Sterilisation of the unfit is long overdue. (Cited in Greer, *Sex and Destiny*, 360)

Indeed, currently a number of private members' Bills being considered in Parliament in 2013 also envisage compulsory sterilization in the interests of the health of the nation.

eugenics of Galton and the negative eugenics of Clapperton. Those who received the attention of the latter were not only criminals, but also the mentally retarded, the insane, the tuberculous, lepers, alcoholics, epiletics, the 'feeble minded', the 'degenerate', immigrants, and of course the poor, who apparently bred all these characteristics.

Darwin, canonized his cousin's ideas, writing, 'We now know, through the admirable labours of Mr. Galton, that genius tends to be inherited.'[27] Given the overwhelming influence of Darwin, it is not surprising that many contemporary commentators began to use Darwinian metaphors to describe social arrangements. Believing that biology was destiny, at least for the poor, Social Darwinists began to identify an extraordinary series of social and behavioural facts as heritable. From this, it was easy to assert that only the 'fit' should be encouraged to procreate and the 'unfit' discouraged.

The IQ test was designed in part to select people deemed eligible for eugenic sterilization. Laws for eugenic sterilization were passed in the USA, the USSR, Canada, and the Scandinavian countries. Eugenics held great appeal for influential people outside of Britain. A prominent eugenist in Germany wrote:

> Because the inferior are always numerically superior to the better, the former would multiply so much faster—if they have the same possibility to survive and reproduce—that the better necessarily would be placed in the background. Therefore a correction has to be made to the advantage of the better. The nature... [*sic*] offers such a correction by exposing the inferior to difficult living conditions which reduce their number.[28]

He added a 'final solution' to the 'correction' offered by nature's lethal ways. Adolf Hitler included in his grand design Jews, communists, homosexuals, and gypsies, among others.

In the USA the eugenics movement gained momentum early in the twentieth century mainly at the instance of natural scientists

---

[27] Charles Darwin cited in Daniel J. Kevles, *In the Name of Eugenics: Genetics and the Uses of Human Heredity* (Cambridge, MA: Harvard University Press, 1995), 20.

[28] Adolf Hitler in *Mein Kampf*, cited in Lars Bondestam and Staffan Bergstrom, eds, *Poverty and Population Control* (London: Academic Press, 1980), 16.

convinced by Galton that 'genius' was a heritable characteristic. In 1906, a Eugenics Section of the American Breeders Association was established to 'emphasise the value of superior blood and the menace to society of inferior blood'.[29] The American eugenic movement involved itself with legislation to restrict immigration of those who were not Anglo-Saxon or Nordic, described as the 'annihilator of our native stock'[30]. It was also instrumental in initiating legislation to carry out sterilizations on institutionalized mentally 'sub-normal', the epileptic, and the psychotic.[31]

An admirer of Hitler in India, M.S. Golwalkar, the founder of the Rashtriya Swayamsevak Sangh (RSS),[32] wrote:

[29] Dennis Hodgson, 'The Ideological Origins of the Population Association of America', *Population and Development Review* 17, no. 1 (March 1991).

[30] Had these immigration laws not been tightened for eugenic reasons, many Jews in Germany and Poland might have escaped the death camps. The Nazi sterilization laws were largely shaped by US Legislation, upheld by the liberal justice. Oliver Wendell Holes as 'it's better for all the world'. See Harry Brunius, *Better for all the World: The Secret History of Forced Sterilisations and America's Quest for Racial Purity* (New York: Alfred A. Knopf, 2006).

[31] Laws for eugenic sterilizations were enacted, other than in the USA, in the USSR, in Canada, and the Scandinavian countries.

[32] The RSS or the Rashtriya Swayamsevak Sangh is the shadowy head of a group of right-wing Hindu organizations collectively called the Sangh Parivar or the Sangh family. The RSS is a male, largely upper-caste, cadre-based organization involved in ideological work and behind-the-scenes-politics. Its Parliamentary wing is the political party the Bharatiya Janata Party (BJP), which headed a coalition government from 1998–2004. The Prime Minister at that time, Mr Vajpayee, is an RSS member, as is the then second-in-command, Mr Advani. Indeed most leaders in the BJP come from RSS ranks. The Sangh family also comprises the more militant, lumpen, Vishwa Hindu Parishad that has played a large role in violence against Muslims and Christians over the past few years, along with another organization named the Bajrang Dal. The women's wing of the RSS is named the Rashtriya Swayam Sevika Sangh and that of the Bajrang Dal is named the Durga Vahini. The Sangh Parivar also has organizations working among students, tribals, workers, and— extremely important for fund-raising—among ex-pat Indians in the USA and the UK, through an organization known as the Hindu Sevak Sangh. Upper caste professional Hindus in these countries, searching for identity, myths, and symbols, are massive funders of fascist organizations in India. There are

To keep up the purity of the Race and its culture, Germany shocked the world by her purging the country of the Semitic Races, the Jews. Race pride at its highest has been manifested here. Germany has also shown how well-nigh impossible it is for races and cultures, having differences going to the root, to be assimilated into one united whole, a good lesson for us in Hindustan to learn and profit by.[33]

Thus, it is not entirely surprising that all fundamentalist agendas include calls for national and racial purification by purging the nation of the Other. Nor is it surprising that those inspired by such claims frequently resort to genocide, targeting in particular women and children. In its less virulent form, the victims of all this eugenic hysteria are the weak, the powerless, and the helpless.

In India, eugenics combined with Victorian utilitarianism in the nineteenth century and British fears generated by the 1857 uprising to prompt a hardening of ideas about race. Indeed, some have argued it was the time when race was created and endlessly classified. The Indian population was categorized into 'martial' and 'non-martial' races; and religion was soon imposed on this classification. Ideas of a Hindu race merged with those of a nascent, aspirational Hindu nationhood; the discovery of an ancient Aryan heritage prompted a eulogizing and harking back to an (imagined) golden period, supported by the work of Indologists such as Max Mueller and others.

Back in London in 1930, the Eugenic Society, encouraged by a Report of the Joint Committee of the Board of Education—which found that 'mentally deficient parents create centers of degeneracy and

---

enormous problems with this characterization of the Sangh Parivar as Hindu fundamentalist or Hindu nationalist. In the first place, they do not represent Hindus, and indeed seem to be deeply ashamed of Hinduism, wishing to transform it into a more 'masculine' religion, such as Christianity or Islam. There are in fact no fundamentals in Hinduism. Their claim to be nationalistic is equally moot since they played an extremely marginal role, if at all, in India's freedom struggle against the British. Indeed, the assassin of Mahatma Gandhi, a good and proper Hindu, was a member of the Sangh Parivar as it then existed. But 'the Hindu fundamentalists' is how they are referred to, especially in Western literature and media, and thus in India.

[33] M.S. Golwalkar, *We, Or Our Nationhood Redefined* (Nagpur: Bharat Publications, 1939), 10.

disease which welfare work can never reach'—began concerted lobbying and propaganda for a Eugenic Sterilisation Bill. Associated with this effort were a press baron, the noted author H.G. Wells, Darwin's son Major Darwin, and Julian Huxley. The last (who later became a stalwart of the population control movement) wrote: 'The principle of supplementing the segregation of defectives by sterilisation in certain cases is to my mind very important, and indeed very essential, if we are to prevent the gradual deterioration of our racial stock.'[34]

Indeed, eugenic ideas had long been tied to the destiny of a nation. A strong nation, especially an imperial one, required more than merely economic and military power; it also needed to ensure its population was kept fresh, energetic, efficient, and productive, preferably by being descended from the 'better stock'. Imperial anxiety about national fitness was exacerbated at the time of the Boer War (1899–1902), when a shockingly low standard of health among male recruits to the British Army, linked together issues of public health and the welfare of the nation. Such anxieties also stoked British fears of German naval strength at the beginning of the twentieth century. After all, American eugenicist Leon F. Whitney could not 'but admire the foresight of the (German) plan (of sterilizing 4,000,000 people) and realize by this action Germany is going to make herself a stronger nation.' (Whitney also observed in his own country that 'the Negroes furnished six times as many sub-normals as did the native-born whites.') In Britain, it was these anxieties and fears that led to the 1904 *Report on the Physical Deterioration of the Population*. It was these concerns that led eugenists Sidney and Beatrice Webb to make their plea for state-sponsored public health in their 1910 treatise, *The State and the Doctor*.[35]

Eventually eugenics was scientifically discredited by biologist (and friend of India) J.B.S. Haldane. But it was Herman Mueller's discovery of genetic mutations in the early 1940s that denuded it of the very last vestiges of scientific respectability. After the Second World War, as revelations of Hitler's gas chambers emerged, the eugenic lobby turned to population control. In 1956, the British Eugenics Society decided in a resolution:

[34] Greer, *Sex and Destiny*, 270.

[35] Ann Oakley, *The Captured Womb: A History of Medical Care of Pregnant Women* (London: Basil Blackwell, 1986).

That the Society should pursue eugenic ends by less obvious means, that is by a policy of crypto-eugenics. The Society's activities in crypto-eugenics should by pursued vigorously, and specifically that the Society should increase its monetary support of the Family Planning Association and the International Planned Parenthood Federation.[36]

OVERPOPULATION AND EUGENICS
IN FAMILY PLANNING IN INDIA

The first family planning clinic in India was opened in 1925 by Karve who went on to formulate official national population policy as a member of the National Sub Committee on Population. The Indian chapter of the Neo-Malthusian League was inaugurated in Madras in 1928.[37] In 1930, the world's first government-sponsored birth control clinic was inaugurated at the behest of the Maharaja of Mysore. Madras soon followed with the establishment of birth control clinics in state hospitals in 1933. In 1935 the Family Hygiene Society was established in Bombay; the Society published *The Journal of Marriage Hygiene*. In the same year, American family planning pioneer Margaret Sanger undertook a nationwide tour of India, winning friends and influencing people (although she left Mahatma Gandhi singularly unimpressed). One of those won over, Lady Dhanvantri Rama Rao, invited her to address the All India Women's Conference.[38]

In 1938 Lady Rama Rao and Margaret Sanger organized the First Family Hygiene Conference in Bombay. They were assisted by an Indian millionaire whose American wife, sharing Sanger's conviction that India was poised at the precipice of a population explosion, had established a Watamull Foundation to control it.[39] The Indian National Congress established a National Planning Committee under the chairmanship of Jawaharlal Nehru. The deliberations of one of its sub-Committees, chaired by Radhakamal Mukherjee, were devoted to the

[36] Cited in Greer, *Sex and Destiny*, 278.

[37] Ashish Bose and P.B. Desai, *Studies in the Social Dynamics of Primary Health Care* (Delhi: Hindustan Publishing, 1983).

[38] Mamata Lakshmanna, *Population Control and Family Planning in India* (Delhi: Discovery Publishing House, 1988).

[39] Greer, *Sex and Destiny*.

question of population, although its concerns were largely eugenic in nature. It deplored the fact that 'attention to eugenics or race culture are matters hardly yet in the public consciousness of this country' and went on to say:

> Man, who has come to the stage of development where he is anxious to breed carefully such species of the lower animals as dogs or horses to obtain very specific qualities in particular specimens of the species, has not yet realised apparently the possibilities inherent in careful scientific breeding of the human race.[40]

It was early in the twentieth century too that what today is called Saffron Demography began to develop.[41] As early as 1909, U.N. Mukherji, in the wake of the partition of Bengal, and colonial initiatives towards communal representation under the Minto Morley reforms, had written *Hindus: A Dying Race*, a book that went on to influence many tracts and publications by the Hindu Mahasabha, the parent organization of the RSS. This book was reprinted many times, simultaneously feeding into and helping to create Hindu communalism. It had a special appeal to those who were anxious to create a monolithic Hindu community, in the face of demands for separate representation emanating from both Muslims and the lower-castes. Whipping up anxiety about Muslims was one way of welding together hugely diverse, and often antagonistic, castes into one community, thereby erasing the structural divisions in caste society.

As the work of historian Sanjam Ahluwalia reveals, the construction of the discourse of overpopulation in colonial India was wedded to the projects of nationalism, modernity, and some varieties of feminisms. The authors of the project were a motley lot: a substantial section of the Indian political elite—with the notable exception of M.K. Gandhi, predominantly male medical practitioners, Western feminists, the early Indian feminist movement, colonial authorities, a section of indigenous

[40] National Planning Committee, *Report of the Sub-Committee on Population* (Bombay: Vora and Co., 1948), 17.

[41] Saffron Demography is how Patricia and Roger Jeffrey describe the myths and lies disseminated by the use of demography by the Sangh Parivar. Saffron is the colour that the Sangh claims represents Hindus. See Jeffery Patricia and Roger Jeffery, *Confronting Saffron Demography: Religion, Fertility and Women's Status in India* (Delhi: Three Essays Collective, 2006).

medical practitioners, and Hindu fundamentalists.[42] Their concerns were equally varied: to harness procreation towards a racially superior country; to hasten economic development; to reduce maternal and child deaths; to decrease the birth rate of the poor, the unfit and the Other, including the lower castes and Muslims; and, finally, to create a modern Indian family, indeed the modern Indian woman married not only to her husband but also the nation, and thus exercising reproductive restraint. Being prudent, a good mother, and god-fearing, she would, above all, be unlike a materialist Western woman.

While Margaret Sanger and Marie Stopes tried to influence the medical profession and the political elites, they made no efforts at all to reach out to either Periyar or Ambedkar, leaders of lower-caste movements who were fierce proponents of both modernism and of birth control, although not for eugenic or neo-Malthusian reasons. Ambedkar, one of those who framed the egalitarian Indian Constitution, called for birth control to liberate women from the tyranny of caste, the family, and patriarchy—a stance too radical for Margaret Sanger. As Ahluwalia notes, 'Even while these Western advocates (of population control) espoused somewhat radical ideas about gender politics, their class affiliations and political programme were clearly middle class. Ambedkar's social base and his radical programme of social transformation made him an unsuitable ally.'[43]

Sanger did try very hard, however, to win over Gandhi, but was rebuffed. Gandhi argued instead, 'If it is contended that birth control is necessary for the nation because of over-population, I dispute the proposition. It has never been proved. In my opinion, by a proper land system, better agriculture and a supplementary industry, this country is capable of supporting twice as many people as there are today.'[44] Gandhi was also opposed to birth control because he believed sex represented sinful indulgence, a lowly animal instinct meant merely for procreation. Ahluwalia observes that 'Gandhi stripped women of sexual desires and reaffirmed motherhood as the natural vocation for women.'[45]

[42] Sanjam Ahluwalia, *Reproductive Restraints: Birth Control in India, 1877–1947* (Ranikhet: Permanent Black, 2008).

[43] Ahluwalia, *Reproductive Restraints*, 63

[44] Ahluwalia, *Reproductive Restraints*, 79

[45] Ahluwalia, *Reproductive Restraints*, 78

Thus middle class women entering nationalist politics were deeply troubled and divided. While they fiercely supported Gandhi's nationalist agenda, they felt strongly about birth control too—for reasons of the health of women and children, among others. Begum Hamid Ali for instance called for 'a safe and really cheap type of contraceptive which may be popularised or given free at clinics established for the purpose'.[46] Kamala Devi Chattopadhyay called for birth control because 'all imperialistic minded rulers encourage large families.... The other reason is because women freed from the penalty of undesired motherhood will deal a death blow to man's vested interest in her. He can no more chain and enslave her.[47] Lakshmi Bai Rajwade, who was to become a forceful advocate of state-provided family planning in the National Sub Committee on Health, argued that child and maternal mortality are directly related to frequent childbirth.

Nevertheless, all these upper-caste women were careful to state that they did not intend contraceptives to rock the familial boat. They too reified the sacred duties of motherhood, even as they strived for it to be carried out in a more rational, prudent manner. To its credit, the Indian women's movement was utterly untouched by communal concerns, but they were nevertheless united by dysgenic fears that the poor were breeding more than the middle classes.

The Second World War temporarily diverted the planners' attention. In 1949, the Family Planning Committee was formed in Bombay with Lady Rama Rao as president. This was renamed as the Family Planning Association of India two years later in 1951. The FPAI has been a major force shaping population policy in India. It takes the credit for 'playing an active role in inducting the first Planning Commission to incorporate family planning in health'.[48] Financial assistance to the FPAI has always been largely provided by international agencies, particularly the Rockefeller Foundation supported International Planned Parenthood Federation (IPPF) and USAID.

During the 1950s the population control lobby, headquartered in the United States, consolidated as a global movement. The American

---

[46] Ahluwalia, *Reproductive Restraints*, 96

[47] Ahluwalia, *Reproductive Restraints*, 105

[48] Family Planning Association of India, *FPAI All India Council, 1973–75* (Bombay, 1975), 1.

millionaire Hugh M. Moore established the Hugh Moore Fund which published a pamphlet, 'The Population Bomb', in 1954. The Ford Foundation joined Moore and Rockefeller in their activities. In 1952, Margaret Sanger and Lady Rama Rao launched the IPPF in Bombay. Invited to this conference was the prominent eugenist C.P. Blacker, who, setting the agenda, noted:

> Nor need we question that a husband and wife living in squalor and ignorance who already have a large number of children not being reared properly might well be considered unfit to have additional children. Yet many parents of these various unfit types keep producing unduly large numbers of children, chiefly because through ignorance or indifference—and often against their will—they let Nature take its course. To combat this situation, eugenists favour the spread of birth control.[49]

The IPPF has been a major force in the population control movement across the world. Funding for the IPPF initially came from the Hugh Moore Fund and Rockefeller Foundation; soon it attracted money from Du Pont Chemicals, Standard Oil, and Shell. Today, on the board of IPPF sit representatives of Du Pont, US Sugar Corporation, General Motors, Chase Manhattan Bank, Newmont Mining, International Nickel, Marconi RCA, Xerox, and Gulf Oil, a veritable Who's Who of America's corporate and finance capital.

The Rockefeller Foundation and the Milbank Memorial Fund founded an Office of Population Research at Princeton University. The Office included leading demographers such as Kingsley Davis and Frank Notestein, who later moved to the Rockefeller founded Population Council.[50] Hugh Moore founded the World Population Emergency Campaign in 1960 with funds from his foundation and from Du Pont. The president of the World Bank ran the campaign whose primary aim was to create and reinforce First World fears of a population 'explosion' in Third World countries with dire consequences for the entire globe.[51] It has been suggested that the communist revolution

---

[49] Cited in Greer, *Sex and Destiny*, 341.

[50] Linda Gordon, *Women's Body, Women's Right* (Harmondsworth: Penguin, 1976).

[51] Susan George, *How the Other Half Dies* (Harmondsworth: Penguin, 1979).

in Cuba provided additional impetus to these fears.[52] Between them, the Emergency Campaign and the Population Council began a systematic and powerful campaign to persuade US policy makers to include population control as a component of US aid to Third World countries. India—always at the forefront of the family planning movement's field of vision—saw Neo-Malthusianism thrust forcefully into its official policy and programmes. They not only funded research but were also involved in training demographers, doctors, and statisticians. It is not surprising that large and influential sections of Indian society today still fervently uphold these Neo-Malthusian ideas.

Over the years, attention to population control increased, even as the state's commitment to health declined. It would not be exaggeration to state that control of population moulded health policies.

NEO–LIBERAL HEALTH

In the late twentieth century, India was still at the heart of global population concerns, especially in relation to China. It was felt that population control had to be successful in India to stave off communism.[53] Elites in India became as convinced as colonials that poverty in the country is primarily due to overpopulation. Neo-Malthusian ideas shaped Indian population policy as it lurched from one approach to another, reaching its apogee during the period of internal emergency in the country between 1975 and 1977, when, officially it was admitted that, 1776 people, many of them men, had died as a result of 'excesses' committed in the family planning programme.[54] Nonetheless, the Prime Minister of India, Indira Gandhi, gained the UN Population Award in 1983. Coercion of various sorts, and all manner of incentives and disincentives continued to characterize the programme, until the entire public health system was suborned under family planning,[55]

[52] Mass, 'An Historical Sketch of the American Population Control Movement'.

[53] Eric B. Ross, *The Malthus Factor: Poverty, Politics and Population in Capitalist Development* (London: Zed Books, 1998).

[54] Rebecca Williams, 'Storming the Citadels of Poverty: Family Planning under the Emergency in India, 1975–1977', *The Journal of Asian Studies* 73, no. 1 (2014).

[55] Mohan Rao, *From Population Control to Reproductive Health: Malthusian Arithmetic* (New Delhi: Sage Publications, 2004).

claiming an increasing proportion of funding relative to that for health. There were a few critical voices, but these few were ignored as the population control juggernaut rolled on.[56]

The government's commitment to public health, never very marked, declined sharply since 1991 when India embarked upon its cost-cutting Structural Adjustment Programmes at the behest of the World Bank and the IMF. India has among the lowest public health expenditure in the world. In 2010, at 1.1 per cent of GDP, it represents the fifth lowest country in the world.[57] More than 70 per cent of health expenditure in the country comes 'Out of Pocket'.[58]

At the same time, policy measures have encouraged the growth of for-profit health care resulting in India having the largest, and least regulated, private health care industry in the world. Evidence from across the country indicates that access to health care has declined sharply. The levying of user fees for public health facilities has been one of factors contributing to the decline in access, especially for poor and marginalized communities and for women.[59] With the sharp rise in health care costs, whether public or private, medical expenditure has become one of the leading causes of indebtedness. Medical expenditure is estimated to send 55 million people into poverty each year.

Since the onset of liberalization policies inter-regional, rural–urban, gender, and economic class differentials in access to health care in India have widened considerably. The decline in public investments was matched by growing subsidies to private sector medical care both for profit and non-profit.[60] There has been a provision of land at

[56] See, for instance the work of Debabar Banerji and Imrana Qadeer cited in Rao, *From Population Control to Reproductive Health*. Since the 1980s, the health and Indian women's movements have critically examined the health and population control programmes in India and have managed to stall the introduction of potentially coercive injectables and implants, besides campaigning against incentives and disincentives in the programme, in particular the Two Child Norm.

[57] WHO, *World Health Statistics* (Geneva: WHO, 2012).

[58] Shiva Kumar, Lincoln C. Chen, Mita Choudhury, Shiban Ganju, Vijay Mahajan, Amarjeet Sinha, and Abhijit Sen, 'Financing Health Care for All: Challenges and Opportunities', *The Lancet* 377, no. 9766 (2011): 668–79.

[59] Government of India, Ministry of Health and Family Welfare, *Report of the Commission on Macroeconomics and Health* (New Delhi, 2005).

[60] Rama Baru, *Private Health in India: Social Characteristics and Trends* (New Delhi: Sage Publications, 1998).

throw-away prices, customs duty exemptions for sophisticated medical technology imports, and loans from financial institutions at low interest rates. As a result, the corporate health sector has become increasingly influential in policy setting. The Confederation of Indian Industry projects health care as a cornucopia of growth, adding enormously to the GDP and to job creation. This sector of the industry, now moving beyond the major metropolitan centres, is said to be flush with private equity funds.[61] One study, for instance, indicated that these incentives had been taken up primarily by urban-based institutions that had not necessarily provided free services to the poor as they had undertaken to do under the terms of the contractual agreement.[62] There has not been enough documentation on the transfer of public health facilities to private providers on a contract basis. One case involved the transfer of ownership of a public tertiary care hospital in Mumbai as part of the state health system project funded by the World Bank.[63] In a recent controversy, a peripheral hospital was also handed over to a private medical college that did not have the necessary facilities and was not recognized by the Medical Council of India.[64] Further, no mechanisms were set up to monitor these projects. A government committee set up to examine the commitments made by private hospitals under the terms of their agreements found gross violations on every undertaking: in short, the public had been diddled, while the government looked benignly on.[65] The Committee's report, however, simply gathers dust.

Public subsidy to the private sector has been achieved within the public health system also. This has been achieved, through the initiation

[61] Indira Chakravarthi, 'The Emerging "Health Care Industry" in India: A Public Health Perspective', *Social Change* 43, no. 2 (2013): 165–76.

[62] Ramesh Bhat, 'Private Health Care Sector in India: Issues Arising out of Its Growth and the Role of the State in Strengthening Public–Private Interaction', unpublished, IIM, Ahmedabad, 1998.

[63] Sunil Nandraj, V.R. Muraleedharan, Rama V. Baru, Imrana Qadeer, and Ritu Priya, *Private Health Sector in India: Review and Annotated Bibliography* (Mumbai: CEHAT, 2001).

[64] Mohan Rao and Oommen C. Kurian, 'India's Health', in *India: Social Development Report 2012: Minorities at the Margins*, ed. Zoya Hasan and Mushirul Hasan (Delhi: Oxford University Press, 2013).

[65] Government of Delhi, *High Level Committee for Hospitals in Delhi: Enquiry Report* (New Delhi: Justice A.S. Qureshi Committee, 2001).

of private–public partnerships (PPP), which took a variety of forms[66] and had international imprimatur. Globally PPPs emerged as the new mantra for a tired, financially emasculated, and visionless WHO in 'partnership' with ruthlessly energetic, new international NGOs, specifically those targeting HIV/AIDS, tuberculosis, and malaria.[67]

These PPPs provided a new impetus to vertical health care programmes but with a focus on a for-profit approach. For instance, under the aegis of Global Alliance for Vaccines and Immunisation (GAVI) and the WHO, the Ministry of Health and Family Welfare plans to introduce a range of unnecessary vaccines.[68] These are premised on GAVI conditions that the government ensures a guarantee for 'reasonable prices', support for credible and sustainable markets, and a prohibition on compulsory licensing. India has accepted GAVI support for the introduction of a hepatitis B vaccine in selected pilot projects. This leads to the creation of new markets along with the neglect of the determinants of diseases.[69] It also effectively undermines comprehensive public health care: routine immunization rates for diphtheria, pertussis, and tetanus (DPT), for instance, have dropped in several states in the country. The alluring poetry of PPPs is a clear case of siphoning resources away from where they are needed, a process the French call *actione capillaire*. In a recent review covering thirty PPPs in reproductive health services in India, Sundari Ravindran observed that there are few assessments of the contribution of such partnerships. This is particularly true of partnerships that do not involve either the central or state governments.[70]

---

[66] Two states, leading advocates of PPPs, Andhra Pradesh and Gujarat, have entered into PPPs in the health sector with Satyam computers, at considerable cost to the state exchequer. Satyam has recently been revealed as the perpetrator India's biggest corporate fraud, its CEO the Madoff of India. The PPPs, however, continue.

[67] Missoni, von Eduardo, 'A Long Way Back towards Alma Ata', *Bulletin von Medicus Mundi Scheiwz*, no. 111 (February 2009).

[68] Indira Chakravarthy, 'Role of the World Health Organisation', *Economic and Political Weekly* 43, no. 47 (2008): 41–6.

[69] Jacob M. Puliyel and Yennapu Madhavi, 'Vaccines: Policy for Public Good or Private Profit?' *Indian Journal of Medical Research* (2008): 1–3.

[70] T.K. Sundari Ravindran, *Public Private Interactions in Reproductive Health Services in India: A Mapping* (Mumbai: CEHAT, 2011).

The encouragement of for-profit health care has also led to burgeoning high-technology diagnostic centres in urban areas with excess capacities. Government employees can now be reimbursed for their expenditure at these institutions, creating effective demand for high cost, high tech medical care. Despite such centres not being warranted by public health considerations, the irrational overuse of their technologies is now commonplace, with Say's Law coming into play: supply in the health sector is over-determined and thereby creates demand.[71]

The oversupply of doctors in the private sector has also led to unnecessary medication of healthy people.[72] Given its higher and better positive emoluments, the private sector also sucks out personnel from the public health system, contributing to significant internal migration to the cities and urban areas. Policy-led development of health tourism has also contributed to this.[73] Thus, while India trains more than enough doctors—but not enough nurses—for its public health system, it is in fact facing an acute staffing crisis in the public health system.[74] Over the same period there has been an increase in private medical colleges charging astonishing capitation fees, which has contributed to appalling standards of medical care.[75] It has also contributed incidentally to skyrocketing dowry rates.[76]

Although India is experiencing this huge crisis of human resources in the public health sector, despite the overproduction of doctors, the

[71] A. Khanna, 'Diabolical Diagnosis', *Tehelka*, 28 August 2004, http://archive.tehelka.com/stoty_mains5.asp?filename-Ne082804Diabolical_Diagnosis.asp (accessed 6 November 2013).

[72] Sunil Nandraj, 'Beyond the Law and the Lord: Quality of Private Health Care', *Economic and Political Weekly* 29, no. 27 (1994): 1680–5.

[73] There is believed to be a flourishing trade in body parts, including reproductive ones in the private sector in India. Newspapers frequently report kidney removal scandals in this black market.

[74] Mohan Rao, Krishna D. Rao, A.K. Shiva Kumar, Mirai Chatterjee, and Thiagarajan Sundararaman, 'India's Health Resource Crises: Too Many and Yet Too Few', *Lancet* 377 (2011), doi:10.1016/S0140-6736(10)61888-0.

[75] N. Ananthakrishnan, 'Medical Education in India: Is It Possible to Reverse the Downhill Trend?' *National Medical Journal of India* 23, no. 3 (2010).

[76] M.N. Srinivas, 'Changing Values in India Today', *Economic and Political Weekly* 28, no. 19 (1993): 933–8.

country is now a major exporter of trained health personnel.[77] The loss of skilled workers usually brings some benefits to the migrants and their families and adds to foreign exchange flows for the country of origin. But it also places a significant strain on the public provisioning and public financing of education. It is estimated, for instance, that 4,000 to 5,000 doctors, trained at public expense, emigrate every year, at an estimated cost of 160 million US dollars to the Indian exchequer.[78] A 2008 study carried out at India's premier public medical institution, the AIIMS in New Delhi, estimated that between 1989 and 2000, half of all doctors trained in India either went overseas or moved internally to the private sector. This study also showed that privileged upper-caste doctors tended to migrate abroad far more than other castes.[79] This of course matches the overall trend over the past few decades for most migrants to the West to be drawn from upper-caste professionals. This trend suggests that affirmative action, reservation of seats in medical colleges, for backward castes has important public health implications.[80]

[77] There is a great deal of anecdotal evidence of unemployment among doctors in India, concentrated in urban areas, who fiercely compete for practices with little attention to medical norms or rational prescription practices. Reflecting this, it appears that doctors are the second largest category of applicants to the Indian Administrative Services exams. Data from the government sector in India, however, is unreliable, while there is little data on the private sector, which is even more averse to academic enquiries. Thus there is no recent data on the private sector in health care in India, leave alone systematic epidemiological data.

[78] Voluntary Health Association of India, *Report of the Independent Commission on Health in India* (New Delhi, 1997).

[79] Manas Kaushik, Abhishek Jaiswal, Naseem Shah, and Ajay Mahal, 'High-end Physician Migration from India', *Bulletin of the World Health Organization* 86, no. 1 (2008).

[80] The question of affirmative action for the backward castes in India is a hugely divisive political issue, playing on familiar tropes from racist and eugenic debates in the West with peculiarly Indian complexities. Doctors and the dominant section of the media, largely upper caste, have opposed reservations advocating that places should be awarded on merit so as to maintain standards. They implicitly argue that characteristics such as intelligence, merit, and efficiency have a genetic basis. Protesting against reservations, medical students have swept streets and polished shoes, considered quintessentially

Initiatives such as the Global Commission on Macroeconomics and Health, presided over by US economist Jeffrey Sachs, spawned similar commissions in countries like India. They have repeated the global commissions' ideas of governance, copying Sach's shock therapy schemes implemented in Russia that led to unprecedented rises in adult mortality.[81] Such 'therapy' has been pungently but accurately described as 'neither public health nor macroeconomics'.[82] While state institutions are criticized for their failure to deliver, no explanation is offered as to why this might be the case or why the private sector is supposed to work better. There is a prevalent assumption in international policy and economic circles that India is characterized by widespread state presence in all sectors of the economy and polity. In the health care sector, however, this is simply not the case. The combination of a weak state sector and an unregulated and powerful private health care sector make it impossible to provide universal care, comprehensive care or above all, equitable care. Tinkering with the combination risks consolidating the dual health care system that India now possesses: one sector weak and underfunded for the vast majority of the population who have little or no access to primary health care, and the other a largely urban-based curative high technology sector for a minority of the population whose public health needs have already been taken care of. Larger macro-economic changes that have widened regional, rural–urban, and class inequalities only compound the

---

lower caste occupations. See Abhay Mishra, 'Anti-Quota Protests: Complaints against Students', *Indian Express*, 8 May 2006.

[81] If institutions now pushing the idea of governance applied it to themselves, the WHO's Commission on Macroeconomics and Health would never have come into being, or Sachs appointed its head, after the health disaster his recommendations precipitated in Russia. But issues of accountability and governance are always selectively raised. Utsa Patnaik has written angrily that economist Amartya Sen, who wrote about Chinese famine deaths, vastly exaggerating the mortality, has not written about the deaths that Sach's measures imposed in Russia. See 'Economic and Demographic Collapse in Russia', in Utsa Patnaik, *The Republic of Hunger and Other Essays* (Gurgaon: Three Essays Collective, 2007).

[82] D. Narayana, 'A Review of Health and Economics', *Economic and Political Weekly* 41, no. 11 (2006): 960–4.

problem. Despite India's economic growth over the last two decades, being invariably described as shining, the country's dim position on the internationally comparative Human Development Index (HDI) reveals that India has the 134th position out of a total of 187 countries.[83] Among the 47 countries in the Medium Human Development group, India is eighth from the bottom.

INDIA'S INTERNATIONAL REPRODUCTIVE TOURISM INDUSTRY: OFFSHORING FERTILITY

The state support for private sector health care, in the context of economic liberalization has spawned India's integration into the global medical tourism industry. This has been policy-led since the last ten years. A subset of this industry is the 'ART' industry, estimated to be worth three billion in the US alone.[84] The Internet is awash with advertisements, enticing couples to come to India for infertility treatment and indeed for surrogacy, with a holiday to the Taj or to Goa thrown in for further allure. Indeed a consortium, INSTAR (Indian Society of Third Party Assisted Reproduction), planned to announce its triumphant arrival with a Surrogacy Walk in the capital, New Delhi, on 20 April 2014. More than 500 surrogate mothers, third party administrators, lawyers, doctors, and grateful parents were expected to participate.[85] 'Companies' offering surrogacy are listed on the stock market.

Within four years of the birth of the world's first test tube baby in 1978, the Government of India sponsored work on IVF at the Institute for Research in Reproduction in Mumbai. The ostensible justification

[83] UNDP, 'Sustainability and Equity: A Better Future for All', *United Nations Development Programme*, 2011, http://hdr.undp.org/sites/default/files/reports/271/hdr_2011_en_complete.pdf (accessed 8 April 2014).

[84] Deborah L. Spar, *The Baby Business: How Money, Science and Politics Drive the Commerce of Conception* (Boston: Harvard Business School Press, 2006). This figure should be treated with caution, because the figure is constantly inflated by those wishing the industry to grow and to flourish.

[85] INSTAR organises Surrogacy Workshop and Walk to Spread Awareness on Third Party Parenting, www.digitaljournal.com/pr/1826048 (accessed 7 April 2014).

was a desire to strengthen the success of the population control programme: since child survival could not be guaranteed in India, the
possibility of creating embryos in the laboratory and then freezing
them might motivate couples to accept sterilization despite their reluctance because of the fear of child's death after sterilization.[86] The IVF
initiative was soon taken over by the booming private health sector,
especially the corporate sector which was beginning to emerge with
state assistance. According to the Indian Council of Medical Research
(ICMR), India had an estimated 250 IVF clinics by 2005 (one of the
highest tallies of any country, if not the highest). The Indian Society
for Assisted Reproduction now has a membership of more than 600.
IVF clinics are now moving in to smaller cities and towns to create and
exploit the market in these areas.[87] A senior official at the ICMR has
estimated that the revenue generated by IVF and related technologies
has leapt five-fold from INR 25,000 crore in 2002 to INR 125,000
crore in 2012.[88] The number of clinics offering IVF also multiplied
five times over the same period.[89]

India has now emerged as a prime destination for people outside
the country who wish to buy some form of assisted reproduction,
either for themselves or for a surrogate. Indeed, India is now described

[86] Sandhya Srinivasan, 'Selling the Parenthood Dream', in *The Unheard
Scream: Reproductive Health and Women's Lives in India*, ed. Mohan Rao (New
Delhi: Kali for Women/Zubaan, 2004).

[87] Sama, *Constructing Conceptions: The Mapping of Assisted Reproductive
Technologies in India* (New Delhi, 2010).

[88] *Economic Times*, 'Surrogacy a $445 Million Business in India', Delhi, 25
August 2008. These figures should be treated with caution since they emanate
from sources with an interest in attracting investments to this area, and thus in
exaggerating both the size of the industry and its possible returns.

[89] India has more than 1,000 clinics in the country offering IVF services.
Estimates suggest that only about one-tenth of them have qualified embryologists. This has given rise to roving embryologists and IVF/surrogacy camps, in
which an embryologist flies in and offers his/her services to several women, all
prepared for embryo implantation, at the same time. At some of these camps,
as many as twenty women have embryos implanted in their wombs in one
session. See V. Deepa, Mohan Rao, Rama Baru, Ramila Bisht, N. Sarojini,
and Susan Fairly Murray, *Sourcing Surrogates: Actors, Agencies and Networks* (New
Delhi: Zubaan Publishing Services, 2013).

as the surrogacy capital of the world, its business worth 445 million dollars.[90] The cost of hiring a surrogate in India ranges from US$6,000 to US$8,000, as against ten times of that, US$80,000, in the USA. The cost of IVF itself in India is about US$500 for each cycle (although there are wide variations), compared to US$5,000 in the USA. Moreover, the ART industry provides a regular supply of leftover ova to the stem cell therapy industry, which is also unregulated in India. This research has corporate and international approval. India has announced a PPP with three European pharmaceutical companies and the British government to carry out stem cell research.[91]

The growth of reproductive tourism to India is justified by its proponents—the government, the assisted reproduction industry, and the middle-men it employs—as a win–win situation: women from abroad, desperate to bear or raise babies with a genetic connection to themselves or their partner, can do so while Indian women earn money as surrogates. But given the highly unregulated nature of medical care in the country, many unethical practices are involved, and ICMR Guidelines governing ARTs are being implemented more in the breach than otherwise.[92] Studies have pointed to the vulnerabilities of the Indian women, driven as they are by quotidian economic concerns to offer their bodies, their fertility, and even their offspring up for exploitation,[93] in a process of servitude if not slavery.[94] Surrogacy is now an international industry, involving a network of actors and agencies, from the global to the local. Many middle-men or third party administrators have emerged to facilitate the process, many of whom have set up NGOs working in rural areas and urban slums to recruit poor women

---

[90] J. Warner, 'Outsourced Wombs', *New York Times*, New York, 3 January 2008. This is a significant amount, but pales in comparison with the trillion dollar global bioeconomy to which the IVF industry offers its extracted raw material that it does not want or need to create a particular baby.

[91] *Hindu*, editorial, Chennai, 23 October 2007.

[92] Sama, *Constructing Conceptions*.

[93] Amrita Pande, 'Not and "Angel", Not a "Whore": Surrogates as "Dirty" Workers in India', *Indian Journal of Gender Studies* 16 (2009): 141–73.

[94] Varada Madge, 'Surrogacy in India: A Case Study of a Surrogacy Clinic in Anand, Gujarat', unpublished PhD thesis, Centre of Social Medicine and Community Health, Jawaharlal Nehru University, New Delhi, 2013.

for surrogacy.[95] The government hesitantly stepped in with legislation to regulate this booming market, but the ART (Regulation) Bill 2010, drafted at the behest of the very industry it seeks to regulate, is meant not so much to offer protection to the women surrogates as to create an aura of responsibility and respectability around the industry.[96] By the same logic, the government could 'regulate' rather than ban sale of body parts and organs, but opted for prohibition in this case.[97] One justification for reproductive tourism is 'reproductive choice' it offers the surrogates, a framework that pays no attention to reproductive or economic justice.[98]

As India integrates further into the global neo-liberal economy it strives to assert itself on the global stage and boasts of an impressive economic growth rate and its ability to withstand global economic meltdown. One way it does so appears to be reproductive tourism and the foreign exchange it generates. The fertility of many Indian women has become a global commodity, as long as the babies to whom they give birth to are those of and for wealthier and/or whiter people—eugenic, racial, caste, and economic logics still operate.

Women, hitherto considered surplus or waste, whose numbers needed to be controlled, are now to be encouraged to have children, albeit for the global reproductive tourist, converting waste into gold. As the French have it, *plus ça change, plus c'est la même chose*. India rushes into globalization, by offering sexual and reproductive slavery as globalized commodities.

[95] Deepa V. et al., 'Sourcing Surrogates: Actors, Agencies and Networks'.

[96] Imrana Qadeer and Mary John, 'The Business and Ethics of Surrogacy', *Economic and Political Weekly* 44, no. 2 (2009): 10–12.

[97] The Transplantation of Human Organs (Amendment) Bill, 2009, that bans commercial organ donations and transplantations has not been unsuccessful in India. But although it might have driven the business underground, it has also vastly reduced the scale of transplants.

[98] Fried Marlene, 'The Politics of Abortion: A Note', in *Markets and Malthus: Population, Gender and Health in Neo-liberal Times*, ed. Mohan Rao and Sarah Sexton (New Delhi: Sage Publications, 2010).

8 Biotechnology in India

Catalyst for a Knowledge Era?

Priya Ranjan

We missed the industrial revolution but we should not miss the information and knowledge revolution.... Leap-frogging into [the] knowledge era looks eminently possible today for our societal transformation in the twenty-first century, which is going to be the century of hope for India.[1]

In 2001, a task force constituted under India's Planning Commission submitted its report recommending strategies to position the country in the global knowledge economy. As is evident in the mentioned quote, the report attributed the information technology (IT) and biotechnology (BT) 'revolutions' with a capacity to help India *leapfrog* into the future. This attribution is not novel. The envisioning of science and technology by Indian planners and policy makers as a catalyst for social transformation predates the 'knowledge era'. The post-colonial developmental model has projected science and technology as one of the main factors of India's social transformation. Typically, this approach took the low level of productive forces as the main cause of underdevelopment in India and identified the import of modern

[1] Government of India (GOI), Planning Commission, 'India as Knowledge Superpower: Strategy for Transformation Task Force Report' (New Delhi, 2001).

technology as a means to 'catch up' with leading economies.[2] India packaged the import of science and technology as a means to solve problems of poverty, food grain production, health, and unemployment. 'Social uplift' and 'technological uplift' became synonymous. Scientific experts, policy makers, planners, and international 'aid' providers depicted technology as a *thing* to be implanted in Indian society. Successive governments have untiringly pushed technological fixes for issues that necessitated more fundamental transformation of social relations.[3] Despite the repeated failures of the Indian state in addressing poverty, unemployment, malnutrition, housing, health, and education, there is a stubborn persistence of this techno-centric approach.

The introduction of neoliberal reforms in India has, however, radically transformed this discourse on development, welfare, and social justice. Advocates of neoliberalism have created a new rhetoric of economic growth fostered by the policies of free trade, privatization, and liberalization. They argue that economic growth would automatically address the issues of poverty and inequality. Within this neoliberal paradigm of economic growth, Indian planners have placed high premium on 'knowledge-based sectors'.[4]

---

[2] R.S. Rao, *Towards Understanding Semi Feudal Semi Colonial Society*, ed. D. Narasimha Reddy (Hyderabad: Perspectives, 1995); Susan George, *How the Other Half Dies: The Real Reasons for World Hunger* (Harmondsworth: Penguin Books, 1991).

[3] The technological fix is not unique to India; it has played a wider role in governmental policies across the world. In 1966, Weinberg, who is said to be the first to use the term, held the view that although only temporarily effective, the technological fix is easier to accomplish than what he termed 'social engineering'. He writes, 'The Technological Fix accepts man's intrinsic shortcomings and circumvents them or capitalizes on them for socially useful ends. The Fix is, therefore, eminently practical and, in the short term, relatively effective', although he admits that 'technological solutions to social problems tend to be incomplete and metastable, to replace one social problem with another'. Alvin M. Weinberg, 'Can Technology Replace Social Engineering?' in *Controlling Technology: Contemporary Issues*, ed. W.B. Thompson (Buffalo, New York: Prometheus Books, 1991), 47; Linda L. Layne, 'The Cultural Fix: An Anthropological Contribution to Science and Technology Studies', *Science, Technology and Human Values* 25, no. 4 (2000): 492–519.

[4] GOI, 'India as Knowledge Superpower'.

The neoliberal paradigm stipulates that the state ought to withdraw from productive activities and welfare measures and allow the market to freely shape every aspect of the economy. This is usually understood as a 'withdrawal of the state'. But this is a generalized rhetoric; by dramatically oversimplifying the complexities emerging through—and brought about by—neoliberal capitalism, it stands in the way of visualizing what exactly this freed neoliberal state goes on to do. Advocates of neoliberal reforms call for the state to concentrate on maintaining 'law and order' and providing the policy and institutional frameworks necessary for the market to operate without constraints. The neoliberal state is thus cast in the role of a 'market activist'.[5] Shedding its pretence of neutrality, it becomes an open promoter and defender of the interests of capital.[6]

Can we agree with the claims of Indian planners and policy makers, actively pursuing neoliberal policies, that IT and BT are harbingers of social transformation? Are these new technologies on their own going to effectively address the problems of poverty, hunger, unemployment, and health? This chapter lays bare the hollowness of the claims of Indian planners and policy makers of revolutionary potential of 'knowledge-based' sectors by examining India's BT-related policies and programmes. It reveals that the Indian BT 'revolution'—the latest instance of India's technological embrace—is intrinsically linked to processes enhancing the interests of local and global capital. It shows that the frog shall only leap in the manner it already knows. The Indian state justifies its support of the biotech industry in the form of investments in research and development (R&D), infrastructural and human resource development, public–private partnerships (PPPs), and subsidies and tax incentives to biotech companies, by using the rhetoric of societal benefits of BT in the areas ranging from food production to health. Contrary to this rhetoric, the chapter asserts that BT is squarely

---

[5] Lave, Mirowski, and Randalls write, 'Neoliberalism diverged from classical political liberalism by renouncing the passive notion of a laissez-faire economy in favor of an activist approach to the spread and promotion of "free markets."' Rebecca Lave, Philip Mirowski, and Samuel Randalls, 'Introduction: STS and Neoliberal Science', *Social Studies of Science* 40, no. 5 (2010): 661.

[6] Prabhat Patnaik, *Re-envisioning Socialism* (New Delhi: Tulika Books, 2011), 87–8.

rooted in capital's logic of accumulation and profit maximization and that its real benefit accrues not to people but to the biotech industry. It is not as though India's promotion of corporate interests is unique to the field of BT. Similar processes are at work in other sectors such as IT and mining. Through an analysis of India's promotion of BT sector, this chapter unravels the workings of a neoliberal state. One of the central tasks of the neoliberal state is to create institutional and policy frameworks necessary for the promotion of the 'market'. It also invests in the development of an infrastructural base crucial for the growth of the private sector. The government's promotion of BT in India exemplifies these characteristics of a neoliberal state. There is a clear division of role and responsibility between the state and industry in BT sector in India. The activities of the Indian government have focused primarily on human resource and infrastructure development and R&D, whereas biotech companies have been involved in the development, manufacture, and sale of biotech products and services. Policy makers and planners rationalize state's investment in BT sector by arguing that it will not only contribute to India's economic growth but also address a range of social issues. The stated vision of the Department of Biotechnology (DBT) during the Eleventh Five Year Plan (2007–12) included the creation '… of biotechnology tools and technologies that address the problems of agriculture productivity, food production, nutrition security, health care and environmental sustainability by providing new and emerging technology, products and services at affordable prices ….'[7] Despite such claims, government documents are unable to provide substantial evidence on the nature and extent of BT's contribution towards addressing the issues of food production, malnutrition, and health. What is evident, however, is the increasing revenues of biotech companies.

BIOTECHNOLOGY AS INDIA'S FUTURE

It is important to note that the modern BT sector has its own specificities, which is intrinsically linked to the ascendency of neoliberalism in the last four decades. We need to foreground India's promotion of BT

---

[7] Cited in GOI, 'Report of the Working Group for the Twelfth Five Year Plan, Department of Biotechnology (2012–2017)' (New Delhi, 2011), 118.

sector in this larger global context in which finance capital with the active support of states is nurturing BT as an important area for capital accumulation. The 1970s' global economic crisis and the attendant hegemonic rise of finance capital in the form of neoliberal capitalism saw the identification of BT as a new field where overaccumulated capital could be invested for further surplus generation.[8] The economic crisis of the 1970s, which was a crisis of over-accumulation, had substantially reduced profitable investment opportunities in the leading capitalist economies, particularly the US. At the same time, a sharp increase in the real interest rates had led to an exponential rise in the wealth in the hands of financial institutions. This led to a desperate hunt for new areas of investment by financial institutions, as rising inflation threatened to devalue idle capital. It was at this moment that big pharmaceutical companies and venture capitalists identified the emerging field of BT as one of the most promising areas for investment.[9]

New developments in life sciences during the 1970s and 1980s created hope as well as hype for a future where many health, food, and environment-related problems would be solved using emerging technologies. Visions of future benefits from biotechnological innovations coalesced perfectly with the modus operandi of finance capital, which creates value not just through investment in productive activities but also through speculations and futures trading. In order to attract investments and to maximize the returns to their investors, biotech companies create a vision of promissory future products.[10] Speculative investment in BT creates 'value in the present to make a certain kind of future possible, [for which] a vision of that future has to be sold, even if it is a vision that will never be realised'.[11]

The promissory character of BT amply aids the neoliberal thrust on states to actively create an investment friendly environment. Important political figures, planners, scientists, and policy makers in India routinely

---

[8] Melinda Cooper, *Life as Surplus: Biotechnology and Capitalism in the Neoliberal Era* (Seattle: University of Washington Press, 2008).

[9] Rodney Loeppky, 'History, Technology, and the Capitalist State: The Comparative Political Economy of Biotechnology and Genomics', *Review of International Political Economy* 12, no. 2 (2005): 264–86.

[10] Kaushik Sunder Rajan, *Biocapital: The Constitution of Postgenomic Life* (Durham and London: Duke University Press, 2006).

[11] Rajan, *Biocapital*, 116.

promote the sector by invoking its benefits to people in the future. Let us consider one such case. In 2003, the then Indian Prime Minister, Atal Bihari Vajpayee, made two rather well known statements on BT in India.[12] In the first statement, he identified 'biotechnology (as) a frontier science with a high promise for the welfare of humanity'.[13] In the second, he expanded the IT and BT acronyms as 'India Today' and 'Bharat Tomorrow' respectively. Here, the acronyms come to signify the Indian 'success' story in information and computational technologies and BT's potential to replicate the same.[14]

Two years earlier, in 2001, Prime Minister Vajpayee had released *Biotechnology: A Vision—Ten Year Perspective*, a 'vision statement' on BT. It outlined measures for 'attaining new heights in biotechnology research, shaping biotechnology into a premier precision tool of the future for creation of wealth and assuring social justice—especially for the welfare of the poor'.[15] Statements promoting and popularising BT as a significant contributor to India's economic growth, as an effective catalyst in addressing issues such as health, poverty, and malnutrition have come not only from India's important political figures, but also from scientists, bureaucrats, and others.[16] Official plan documents,

[12] The year 2003 also witnessed the launch of the Association of Biotechnology Led Enterprises (ABLE), an association of industrialists from India's BT sector. Within ten years, this lobby's membership figures crossed 240 and included representatives from all fields of BT. The association's objectives—to 'influence the government for optimal biotechnology policies and for creation of a clear regulatory environment that promotes innovation [in the biotech field] … promote industry-academia linkages … [and act as a conduit for collaborations] among domestic players, between domestic and international firms'—are in tune with the neoliberal paradigm. Details of ABLE's aims are from http://ableindia.in/Bio_survey_2012_updated.pdf (accessed September 2012).

[13] Nikolas Rose, *The Politics of Life Itself: Biomedicine, Power, and Subjectivity in the Twenty-First Century* (Princeton, NJ: Princeton University Press, 2007).

[14] Aditya Bharadwaj and Peter Glasner, *Local Cells, Global Science: The Rise of Embryonic Stem Cell Research in India* (New York: Routledge, 2009), 30.

[15] Available at the Department of Biotechnology's website http://dbtindia.nic.in/uniquepage.asp?id_pk=102 (accessed December 2011).

[16] Manju Sharma, 'India: Biotechnology Research and Development', in *Agriculture, Biotechnology and Poor: Proceedings of an International Conference, Washington, D.C., 21–22 October 1999*, ed. G.J. Parsley and M.M. Lantin

speeches, reports, and bytes on BT in India resonate with claims and predictions along similar lines. Thus, efforts on the part of actors from a number of fields converge to create social imaginations of BT as a sector brimming with hope and promises.

Let us take the case of GM crops. In the dominant representations of agricultural BT, we see a clear-cut articulation of the technological fix. GM crops are promoted as a way out of agrarian crises, poverty, and environmental issues. For instance, scientists at the Department of BT sought to inscribe 'Bt cotton (as) an example of how timely intro-duction of new technology can break productivity barriers and help crop production in a sustainable manner'.[17] Likewise, according to a scientist at the Indian Institute of Science, one of the country's premier research institutes, India's need to embrace agricultural BT stems from

[its] ever-increasing population ... [its] availability of land [which] is forever shrinking ... [and because] the so-called adequate levels of food grains at present are, perhaps, due to the segment below the poverty line, not being able to access adequate food. If everyone can afford to eat his/her full quota of food, the production may actually fall short of demand.[18]

While the above-mentioned account shares with India's promo-tion of the green revolution in the 1960s and 1970s the bugbear of overpopulation,[19] the addition of supporting arguments is noteworthy. These resonate with the contemporary promotion of agricultural BT in developing countries, which makes a 'moral case for GM crops to feed the hungry and aid "development" in the South'.[20] Moreover, the association of advancements in BT with an all-embracing progress in

---

(Washington: Consultative group on International Agricultural Research, 2000); G. Padmanaban, 'Growth of Biotechnology in India', *Current Science* 85, no. 6 (2003): 712–19; S. Natesh and M.K. Bhan, 'Biotechnology Sector in India: Strengths, Limitations, Remedies and Outlook', *Current Science* 97, no. 2 (2009): 157–69.

[17] Natesh and Bhan, 'Biotechnology Sector in India', 162.

[18] Padmanaban, 'Growth of Biotechnology in India', 714.

[19] George, *How the Other Half Dies*.

[20] Sally Brooks, 'Biotechnology and the Politics of Truth: From the Green Revolution to an Evergreen Revolution', *Sociologia Ruralis* 45, no. 4 (2005): 360–79.

human history and universal human welfare has become all the more necessary with the rising oppositions to GM food crops.[21] The image of an India in the throes of a BT revolution is created not only by political leaders and scientists but also by industry representatives and market analysts. Under neoliberalism, speculation and market prediction appear to play a far greater role in deciding the flows of capital investment than they did previously. Trend predictions on the Indian BT sector have ranged from forecasting India as one of the most favoured destination for contract research and stem cell research to delineating it as one of the strongest players in 'emerging markets' in the Asia-Pacific business region.[22]

Statistics collated by government or industrial associations and estimates generated by market analysis firms show a growing BT sector in India. In 2003, ABLE-*BioSpectrum* predicted that the industry would grow by 26 per cent in 2003–4.[23] By then, established market analyst groups such as Ernst & Young had predicted that the industry's revenues in India would touch US$5 billion and generate one million jobs over the next five years (2003–8).[24] Revenues finally managed to touch

[21] Scepticisms over and oppositions to BT have also led to a heightened astuteness in market analysis. Business analysis of BT markets seeks not only to identify existing oppositions but also to predict any *potential* ones. International bodies such as the OECD expect member countries to incorporate statistical information on '*social attitudes* towards biotechnology' while they collate BT statistics. Sachin Chaturvedi, 'Developments in Biotechnology: International Initiatives, Status in India and Agenda before Developing Countries', *Science Technology & Society* 8, no. 1 (2003): 73–100 (emphasis added).

[22] Ernst and Young, 'Beyond Borders: Global Biotechnology Report 2009', *Biotechnology Journal* 4 (2009): 1108–10, http://www.massey. ac.nz/~ychisti/E&Y09.pdf; Exim Bank of India, 'Biotechnology Industry in India: Opportunities for Growth', Occasional Paper no. 137, 2010, http:// docslide.us/documents/biotechnology-industry-in-india-opportunies-for-growth.html.

[23] Soon after its establishment in 2003, the industrialists association ABLE began to annually conduct surveys of the Indian biotech industry. The results of these are published in the magazine *BioSpectrum* (accessible at http://www. biospectrumindia.com/). The annual surveys can also be accessed at http:// ableindia.in/able_biospectrum_surveys.php. *BioSpectrum*, 'Biotech Industry 2002–03: A Beginning' (September 2003): 28.

[24] *BioSpectrum*, 'Biotech Industry 2002–03', 26.

US$4 billion in 2011.[25] The mismatch between the predicted figures and the revenue figures did not subdue the industry's predictions of a bright future. Instead, it was followed by another round of market forecasts—that annual revenues shall touch US$100 billion by 2025.[26] Statistical figures such as these, along with concepts such as CAGRs, market forecasts, and market worth, also play a part in shoring up the image of 'superpower India' all set to become a world leader in BT.

PUBLIC INVESTMENT, PRIVATE PROFIT

Since its emergence, the global biotech industry has been dominated by US-based companies.[27] The hold of the US over the industry is such that even other leading capitalist economies such as the UK or Germany are nowhere close to challenging it. Everywhere—even in Europe—governments seem to have acquiesced to the fact that they have lost out to the US in the biotech big game and must therefore initiate a series of measures to catch up.[28]

[25] From their annual surveys, ABLE–*BioSpectrum* calculated a compounded annual growth rate (CAGR) of 29.5 per cent for the five year period between 2002 and 2007, and a CAGR of 14.75 per cent for the subsequent five year period. The total revenues generated by BT changed from INR 18,300 million in 2002–3 to INR 85,410 million in 2006–7 and to around INR 204,400 million in 2011–12. In 2011–12, the revenues share of medical BT and agricultural BT constituted 62 per cent and 15 per cent respectively—*BioSpectrum*, '10 Years of Indian Biotech' (June, 2012): 61–75.

[26] *BioSpectrum*, '10 Years of Indian Biotech', 76.

[27] The indisputable dominance of the US in biotech can be summed up by the following figures—four out the top five biotech companies are US-based, and the US also accounts for nearly two-thirds of global biotech revenues. William Bains, *Venture Capital and the European Biotechnology Industry* (New York: Palgrave Macmillan, 2008).

[28] For a summary of the European Economic Cooperation (EEC) countries' initial BT-related initiatives, such as public funding, technology transfer schemes for commercialization, and measures adopted to make up for the lack of huge venture capital funding (a significant point of divergence from the capital markets which shaped US BT industry) and for interactions between BT companies in the US and Europe, see Mark D. Dibner, 'Biotechnology in Europe', *Science* 232, no. 4756 (1986): 1367–72. See Bains, *Venture Capital*, for an account of venture capital's role in the US biotech industry. US hegemony

If leading capitalist economies are themselves grappling with such anxieties, it is no great surprise that the discourse on BT in India has tended to convene around rough and ready imageries such as losing out, lagging behind, and catching up. India has been rather comfortable identifying itself as a laggard state. Clearly, this is a better option than to identify itself as an economy largely dependent on leading capitalist economies. In some accounts, the differences in the BT sector within the advanced capitalist economies supply readymade lessons to India. One such account compares the development of BT sectors in India and France (which lagged behind other advanced capitalist economies such as the US in the early 1980s but acquired greater success later). The author hypothesizes that 'an examination of the French government's approach to the development of biotechnology as a potential strategy for a "latecomer country" might be interesting for any developing country where the government is the principal actor committed to the development of biotechnology'.[29] She then suggests that India emulate the French strategy so as to successfully transform its scientific competence into industrial competence. Such accounts efface the historical roots and social bases of the development of science and technology.

As a matter of fact, BT's first governmental appearance (in policy documents) dates back to the early 1980s, a rather early start for a state that prefers to identify itself as a laggard. This early start is acknowledged by one commentator as follows—'India is one of the first few countries, among the developing countries, to have recognized the importance of biotechnology as a tool to advance growth

---

over the biotech industry is so overwhelming that the then UK Prime Minister Tony Blair was expressing his wish to see Britain emerge as the European hub of BT, or 'the next wave of the knowledge economy', as late as in 2000 (cited in Rose, *Politics of Life Itself*, 35). See also Loeppky, 'History, Technology, and the Capitalist State', for an analysis of the significant delay in the development of BT in Germany.

[29] Shyama V. Ramani, 'Creating Incentives: A Comparison of Government Strategies in India and France', *Biotechnology and Development Monitor* 26 (1996): 18–21. Citations of this article do not include page numbers as I referred to an unpaginated online edition. This is available at http://www.biotech-monitor. nl/2605.htm (accessed January 2012).

of agricultural and health sectors as early as in 1980s'.[30] The same author identifies the Sixth Five Year Plan covering the years between 1980 and 1985 as the earliest policy document to deal with measures for BT development.[31]

On the part of the Indian state, specific efforts to foster the development of BT were set in motion with the setting up of an apex body, the National BT Board or the NBTB in 1982. The board, chaired by a member of India's Planning Commission, comprised representatives from other public institutions related to science and technology and scientific research such as the Council of Scientific and Industrial Research (CSIR), the Indian Council of Agricultural Research (ICAR), the Indian Council for Medical Research (ICMR), the University Grants Commission (UGC), as well as representatives of departments such as the Department of Science and Technology (DST) and the Department of Atomic Energy (DAE).

The NBTB's mandate was to pinpoint specific areas that were to assume priority within the promotion of BT. By 1983, it had executed its chief mission and submitted a report titled the *Long Term Plan in Biotechnology for India*.[32] The long-term programmes delineated by the NBTB lay the emphasis on boosting the necessary infrastructure for BT. This included the setting up of research institutes and university level academic programmes which would enhance the availability of a skilled labour force proficient in BT R&D.[33] It is noteworthy that from the very beginning India has laid stress on the industry–academe

---

[30] Sachin Chaturvedi, 'Status and Development of Biotechnology in India: An Analytical Overview', RIS Discussion Paper no. RIS-DP 28/2002, Research and Information System for the Non-aligned and Other Developing Countries, New Delhi, 2002, 2.

[31] Chaturvedi, 'Status and Development of Biotechnology in India', 2; see also V.V. Krishna, 'Dynamics in the Sectoral System of Innovation: Indian Experience in Software, Biotechnology and Pharmaceuticals', in *Science, Technology Policy and the Diffusion of Knowledge: Understanding the Dynamics of Innovation Systems in the Asia Pacific*, ed. Tim Turpin and V.V. Krishna (Cheltenham, UK: Edward Elgar Publishing, 2007), 209.

[32] Chaturvedi, 'Status and Development of Biotechnology in India', 3.

[33] S. Visalakshi and G.D. Sandhya, 'R and D in Pharma Industry in Context of Biotech Commercialisation', *Economic and Political Weekly* 35 (2000): 4223–33.

link and on commercialization of BT research, a point we shall return
to later in the chapter.

In 1986, the NBTB was replaced by the DBT, a full-fledged department established under the Ministry of Science and Technology. The
DBT has since become the nodal governmental agency for various
BT-related initiatives. Public investment in the BT sector has increased
over the years. The total plan allocation for the DBT increased from
INR 6,210 million in the Ninth Five Year Plan (1998–2002) to
INR 65,000 million in the Eleventh Five Year Plan (2007–12).[34] The
major share of plan allocation goes to human resource and infrastructure development and R&D. More than 70 per cent of total outlay
earmarked for the DBT by the Eleventh Five Year Plan went into
human resource and infrastructure development, R&D, and support to
autonomous R&D Institutions.[35]

During the previous decade, the DBT's focus on medical BT has
increased. Out of a total of 1,225 projects implemented during the first
four years (2002–6) of the Tenth Five Year Plan, 39 per cent projects
were in the field of agricultural BT, and only 20 per cent were in the

[34] It has to be noted that the major share of Science and Technology plan
allocations to Central Scientific and Socio-Economic Departments goes to
the three strategic fields of defence, space research, and atomic energy. During
the Tenth Five Year Plan (2002–7), the Defence Research and Development
Organization, the Department of Space and the Department of Atomic Energy
together obtained close to 80 per cent of budgetary allocations for Science and
Technology, whereas the DBT's share was only 3.14 per cent (P. Banerjee,
'India: Science & Technology 2008', Report, New Delhi, National Institute
of Science, Technology and Development Studies (NISTADS), CSIR, 2009),
43–4). Since the Seventh Five Year Plan (1985–90), Science and Technology
plan allocations have nearly doubled with each plan period. Science and
Technology plan allocations in India may not have reached the levels that
experts call for, but what is significant is that plan expenditure in sectors such
as agriculture, industry and minerals, and rural development have decreased
steadily (Banerjee, 'India: Science & Technology 2008', 42–3); Government
of India, 'Annual Report 2008–9' (New Delhi, 2009); Government of India,
'Annual Report 2009–10' (New Delhi, 2010).

[35] Details are available at http://planningcommission.nic.in/sectors/sci-
ence.php?sectors=sci (accessed August 2013).

field of medical BT.[36] These figures changed during the Eleventh Five Year Plan. Out of a total of 2,410 projects implemented during its first four years (2007–11), 26 per cent projects were in the field of medical BT, whereas only 22 per cent projects were in the field of agricultural BT.[37] There have also been significant changes in budgetary allocations within the sector. During the Ninth Five Year Plan, only 13 per cent of the DBT's total budget was assigned to medical BT; this increased to 36 per cent in the Tenth Five Year Plan.[38]

From its year of inception in 1986, the DBT has been instrumental in the establishment of research institutes and university level departments for human resource development and enhancement of R&D capacities in the BT sector. For instance, the National Tissue Culture Facility (later named as the National Centre for Cell Sciences) and a BT Information System were established at the department's initiative in 1986.[39] Such efforts have increased over the years. By 2013, thirteen autonomous research institutions and two public sector undertakings had been established under the supervision of DBT.[40] The performance of special research centres is dependent, however, on the availability of a steady stream of researchers who have already obtained necessary levels of competence in life sciences and BT. State efforts to develop 'manpower' have therefore also focussed on the introduction of postgraduate BT courses in universities. In the DBT's second year of functioning, that is, in 1987–8, INR 54 million

---

[36] Government of India, Ministry of Science and Technology, Department of Biotechnology, 'Report of the Working Group for the Eleventh Five Year Plan' (New Delhi, 2006), 1–2.

[37] GOI, 'Report of the Working Group for the Twelfth Five Year Plan', 1–2.

[38] Brian Salter, Melinda Cooper, Amanda Dickins, and Valentina Cardo, 'Stem Cell Science in India: Emerging Economies and the Politics of Globalization', *Regenerative Medicine* 2, no. 1 (2007): 83.

[39] GOI, 'Annual Report 2008–9' (New Delhi, 2009), 120, 138.

[40] For details, see the department's website at http://dbtindia.nic.in (accessed January 2013). Public investment in research institutes and university departments specializing in the life sciences and biotechnology is also channellized through other departments and institutions such as the DST, CSIR, ICAR, ICMR, and the UGC.

or 18 per cent of its total R&D expenditure went into 'manpower development'.[41]

Until the 1980s, biological sciences did not enjoy as much state support in India as other science and technology fields such as physics, mathematics, or engineering.[42] India's subsequent promotion of life sciences and BT research is linked to the new developments in life sciences in the 1970s and 1980s and the projection of BT as a technological base of future societies. The DBT/UGC introduced masters' programmes in BT in educational institutions across the country and doctoral research programmes in a few universities. Postgraduate programmes were introduced in six universities in 1985–6. By 2009–10, a total of 72 universities had come to offer both general as well as specialized BT courses. The DBT has additionally extended its support to 35 colleges offering BT courses at the undergraduate level.[43] Within 20 years of the department's existence, its initiatives have also led to an intake of 300 faculty and researchers at the university level.[44] The Indian state has thus focused on building highly specialized research institutions and creating and upgrading the departments and centres of life sciences and BT in various universities and colleges during the last two decades. Such amplified efforts of the Indian state on human resource, infrastructure development, and R&D have been crucial for the development of the BT industry.

The generation of a skilled labour force through public investment in education is by no means unique to 'frontier' areas of science. Higher educational institutions in general are means by which the 'skilling' of a section of the country's labour force occurs. They are an expression of developments in the organization of production and distribution, one aspect of what Harvey terms 'social infrastructures'.[45] While mechanization and technology have confronted the majority

[41] Sunil Mani, 'Biotechnology Research in India: Implication for Indian Public Sector Enterprises', *Economic and Political Weekly* 25, no. 34 (1990): M-117.

[42] R.D. Vale and K. Dell, 'The Biological Sciences in India: Aiming High for the Future', *The Journal of Cell Biology* 184, no. 3 (2009): 342–53.

[43] GOI, 'Annual Report 2009–10', 23.

[44] Krishna, 'Dynamics in the Sectoral System of Innovation', 210.

[45] David Harvey, *The Limits to Capital* (London; New York: Verso, 2006), 398.

of the working class as an alien force, capitalist development has also called for the creation of a set of skilled workers, technical, and scientific 'experts'. The creation of experts and scientists or the skilled labour force of the knowledge economy requires specific efforts and the imparting of specialized skills through educational institutions. The DBT's efforts on human resource development are thus part of the processes through which to sync the country's 'overall human resource complex' with developments in the new field of BT.[46] These efforts create a social infrastructure that is necessary and conducive for capital accumulation in the field of BT.

Moreover, like all labour in India, the biotech skilled labour force is also relatively 'cheap' or 'cost effective'. Wages in the BT sector exhibit vast differences globally. One website calculates the median starting monthly salary in BT as approximately US$5,000 in the USA, GBP 2,500 in the UK but INR 64,000 in India.[47] It is evident that India's biotech skilled labour force is 'cost effective' or 'cheap'. This has been one of the key factors in the growth of BT in India. In fact, this is an important and well recognized factor in India's promotion of BT. This is clearly spelt out, for instance, by Invest India, a joint venture of the Department of Industrial Policy and Promotion (Ministry of Commerce and Industry), the Federation of Indian Chambers of Commerce and Industry, and state governments, which was established to promote foreign investments in India. It counts low labour costs as one of the main advantages of foreign investment in India's BT sector, stating that 'India offers a low-cost and skilled labour force, which is a key reason for the country attracting outsourced research activity from global biotechnology companies'.[48] Similarly, in an interview published in global market analyst firm Ernst & Young's *Beyond Borders: Global BT Report 2009*, M.K. Bhan, then the Secretary, DBT, stated, 'It is our view that India offers Western companies highly competitive, cost-effective

---

[46] Harvey, *Limits to Capital*, 399.

[47] Sourced from http://www.wageindicator.org/main/salary/Salary-checkers. Figures for USA, UK, and India are taken from http://www.paywizard.org/main/salary/calculator, www.paywizard.co.uk/main/pay/salarycheck, and http://www.paycheck.in/main/salary/salary-checker-1 respectively (accessed December 2013).

[48] Available at http://www.investindia.gov.in/?q=biotechnology-sector (accessed December 2013).

alternatives for research and manufacturing work.'[49] This 'cheap work-force' which is advertised so casually by India to leading capitalist economies is created through programmes routing 'public investment' through the DBT into biotech educational and research institutions.

REFRAMING OF PUBLIC–FUNDED SCIENTIFIC RESEARCH INSTITUTIONS

The hegemonic rise of neoliberal economic orthodoxy has radically transformed the linkages between public-funded research institutions and industries. Public-funded universities and research institutes have become far more tuned with the demands and interests of industries. This is not to say that scientific research institutions functioned in a 'disinterested' manner earlier. However, as opposed to earlier periods, they now exhibit an immediate responsiveness to industries' interests. The establishment of close ties between academia and industry has been particularly crucial for the growth of the BT industry. Compared to other sectors of the knowledge economy, such as IT, the development of a new product or service in the case of BT is a much more capital intensive process, requiring longer periods of research.[50]

A number of scholars have identified certain US legislative measures that made scientific research institutions more industry friendly as one of the important factors for the growth of US BT industry.[51] The Stevenson–Wydler Technology Innovation Act and the Bayh–Dole Act, both passed in 1980, prepared the grounds for close 'interaction' between the US academia and industry. These legislations fostered the transfer of technologies developed in federally funded research institutions to industries for their commercialization and made it mandatory for universities to patent technologies developed by them.

---

[49] Ernst and Young, 'Beyond Borders', 109.

[50] Rajan, *Biocapital.*

[51] Paul Rabinow, *Making PCR: A Story of Biotechnology* (Chicago: University of Chicago Press, 1996); Catherine Waldby and Robert Mitchell, *Tissue Economies: Blood, Organs, and Cell Lines in Late Capitalism* (Durham, NC: Duke University Press, 2006); M.J. Lynskey, 'The Dismantling of Redundant Dichotomies: Biotechnology as an Exemplar of University-Industry Collaboration', *Journal of Commercial Biotechnology* 12, no. 2 (2005): 127–47.

These legislative measures were part and parcel of larger neoliberal reforms initiated by the US since the late 1970s. In the US, neoliberal reforms led to the privatization of higher education and R&D. US spending on higher educational institutions declined gradually in the last four decades. In 2006–7, state contribution to the total budgets of public research universities was close to 20 per cent.[52] In the US, public research institutions primarily depend on corporate investment. US corporations have increasingly abandoned the practice of in-house R&D. They outsource their research and development processes to contract research organizations and universities.[53] While the general tendency of neoliberal reforms is to promote privatization, it adopts different strategies in different institutional and political contexts. In India, for instance, scientific research and development activities are mainly carried out by public-funded universities and research institutes. Although the investment of private sector in R&D activities has increased in recent years, its contribution to total R&D expenditure in India still remains close to 20 per cent.[54] However, the neoliberal policies pursued by India since the early 1990s have initiated a series of reforms in public-funded scientific research institutions which have increasingly become more market friendly and industry oriented.

We can take the case of the CSIR to highlight the neoliberal reframing of scientific research institutions in India. Established in 1942, the CSIR is a body that coordinates and supervises a network of 37 public-funded research laboratories and institutes.[55] The CSIR's mandate was to promote scientific and industrial research which would cater to the requirements of industry. It is important to note that in the pre-reform period the industrial sector in India was largely dominated by Public Sector Undertakings (PSUs). In the early 1980s, 'the public sector

[52] Lave, Mirowski, and Randalls, 'Introduction: STS and Neoliberal Science', 665.

[53] Philip Mirowski and Robert Van Horn, 'The Contract Research Organization and the Commercialization of Scientific Research', *Social Studies of Science* 35, no. 4 (2005): 503–48.

[54] Banerjee, 'India: Science & Technology 2008', 47–8.

[55] Eleven of the CSIR research institutes are in the field of biological sciences; eight are in the chemical sciences; eleven are classified under engineering sciences, five under physical sciences, and two under information sciences. Details are available at http://www.csir.res.in (accessed January 2013).

enjoyed a 3:1 advantage over the private sector in paid-up share capital, and a 4:1 advantage in the value of sales'.[56] However, the CSIR's contributions to PSUs, which were largely dependent on imported technologies, were minimal. The CSIR continuously failed to fulfil its mandate.[57]

With the introduction of neoliberal policies, the process of privatization of PSUs and metamorphosis of the CSIR started. In fact, in 1986, the CSIR Review Committee for the Development of Science and Technology laid the foundation for later changes in the CSIR.[58] This Committee was headed by Abid Hussain, who was then a member of the Planning Commission, and a key figure in the upcoming spate of reforms in sectors such as small scale industry, capital markets, textile, manufacturing, and so on. The Abid Hussain Committee emphasized the CSIR's inability to commercialize scientific research.[59] One of the crucial recommendations of this Committee was to end the monopoly of the National Research Development Corporation (NRDC) over technologies developed by the CSIR's laboratories and research institutes. This allowed CSIR institutes to directly collaborate with the industry.[60] In 1992, the CSIR constituted another committee chaired by R.A. Mashelkar to explore ways and means to make the council more market oriented.[61] In order to promote scientific innovations, the Mashelkar Committee proposed economic incentives for CSIR scientists. According to Sunder Rajan, the Mashelkar Committee's re-envisioning of the CSIR also 'involve[d] generating external, non-federal revenue, increasing annual earnings from overseas R&D, developing licensable technologies (of which there were none in 1994) and obtaining foreign patents and using them to fund operational expenditures'.[62]

[56] Francine R. Frankel, *India's Political Economy: 1947–2004*, 2nd ed. (New Delhi: Oxford University Press, 2005).

[57] Rajan, *Biocapital*, 214.

[58] Gita Piramal, Sumantra Ghoshal, and Sudeep Budhiraja, *World Class in India: A Casebook of Companies in Transformation* (New Delhi, India: Penguin, 2002), 303.

[59] Mani, 'Biotechnology Research in India: Implication for Indian Public Sector Enterprises', *Economic and Political Weekly* 25, no. 34 (1990): M-122.

[60] Banerjee, 'India: Science & Technology 2008', 102.

[61] Piramal, Ghoshal, and Budhiraja, *World Class in India*, 305.

[62] Rajan, *Biocapital*, 214.

Subsequent institutional reforms have led the CSIR research institutes to focus more on patentable research and commercialization of technologies. By the year 2010–11, the CSIR had given license on its 222 patents to various clients and had procured 3,046 foreign and 2,278 Indian patents.[63] Despite its ambition to become a self-financing organization, the CSIR is still mainly dependent on state funding. In the year 2010–11, more than 80 per cent of the CSIR budget came from the state.[64]

The growing importance of knowledge-based sectors in the larger global economy is one of the important factors strengthening the linkages between academia and industry. Intellectual Property Rights (IPR) is at the core of knowledge-based sectors. The neoliberal rhetoric of promoting innovations in educational institutions through incentives is actually aimed at the creation and commercialization of IPR. The US Bayh–Dole Act, mentioned earlier in the chapter, was a defining moment in that direction. In 2008, in an ostensible effort to emulate the 'success story' of the US in BT sector, the DBT drafted a Bill called The Protection and Utilization of Public Funded Intellectual Property (PUPFIP). The main thrust of this Bill, which was passed by the Indian Parliament in 2010, was to create a uniform legal framework for the protection of the Intellectual Property (IP) generated through public-funded research institutions and provide financial incentives to scientists.[65] The DBT's agenda behind this legislation was to further strengthen the ties between academia and industry by facilitating speedy commercialization of technologies developed in public-funded research institutions.[66]

In fact, this legislative effort of the DBT is in continuation with its earlier policy of technology transfer and its commercialization. In 1990, barely four years into its existence as a full-fledged department, the DBT played an instrumental role in the setting up of the Biotech Consortium India Limited (BCIL). A public limited company

[63] Government of India, Council of Scientific and Industrial Research, 'Annual Report 2010–2011' (New Delhi, 2011), xiv.

[64] GOI, CSIR, 'Annual Report 2010–2011', xxv.

[65] Pradeep Srivastava and Sunita Chandra, 'Technology Commercialization: Indian University Perspective', *Journal of Technology Management & Innovation* 7, no. 4 (2012): 121–31.

[66] GOI, 'Annual Report 2009–2010' (New Delhi, 2010), 190.

incorporated under the Companies Act, 1956, the BCIL was to provide 'the linkages amongst research institutions, industry, government and funding institutions, to facilitate accelerated commercialization of biotechnology'.[67] The BCIL was financed by a clutch of pharmaceutical companies such as Ranbaxy Laboratories and Cadila Laboratories as well as financial institutions such as the Industrial Development Bank of India, the Industrial Credit and Investment Corporation of India Bank Limited and the Unit Trust of India.[68] The establishment of the BCIL was a departure from the earlier state policy for technology transfer and commercialization. Before the BCIL came into the existence, the NRDC, which was established in 1953 under the Ministry of Science and Technology, was the only organization with a mandate to commercialize technologies developed in public-funded research institutions.[69]

The DBT and CSIR place high premium on IP protection, commercialization, and technology transfer. No doubt, commercialization of technologies brings financial benefits not only to individual scientists but also to research institutions.[70] But financial gains are not the only or the most important outcome of commercialization of technologies developed in scientific research institutions. The neoliberal thrust on strong ties between academia and industry and speedy commercialization of technologies decisively influences the very content of scientific research. Scientific research institutions increasingly promote research which serves the needs and requirements of private companies.

INVESTMENTS, COLLABORATION, AND PUBLIC–PRIVATE PARTNERSHIP

In India, the state has been the most important actor in promoting BT sector.[71] This stands in contrast with the US, where venture capital

[67] Augustin Maria, Joël Ruet, and Marie Helene Zerah, 'Biotechnology in India', Report (New Delhi: French Embassy in India, 2002), 21.

[68] Maria, Ruet, and Zerah, 'Biotechnology in India', 21.

[69] Banerjee, 'India: Science & Technology 2008', 101–2.

[70] In 2007–8, NRDC earned INR 40.2 million from licence fee and royalty. The BCIL's total earning in 2005–6 was INR 6.8 million (NISTADS 2009: 103–4).

[71] Praveen Arora, 'Healthcare Biotechnology Firms in India: Evolution, Structure and Growth', *Current Science* 89, no. 3 (2005): 458.

investment has played a decisive role in the rapid growth of BT indus-try.[72] The difference has led to analyses of the Indian state's attempts to supposedly counter the lack of venture capital investment in the BT sector through initiatives such as the establishment of BCIL as 'an interesting form of state intervention to counter market failure'.[73] In India, the state not only invests in human resource and infrastructural development and R&D to promote the BT industry but also provides financial assistance to biotech companies through various kinds of PPP programmes.

In 2006, the DBT emphasized the promotion of the PPP model in the Eleventh Five Year Plan (2007–12) for accelerating technological innovation. The DBT's *Report of the Working Group for the Eleventh Five Year Plan* stated that 'there is a need to synergize public sector creativ-ity with private sector management excellence to create world class technologies through a variety of public-private partnership models'.[74] State promotion of the PPP model in BT sector became more pro-nounced when, in 2007, the DBT decided to invest 30 per cent of its budget in PPP schemes.[75]

In fact, steps were already taken in this direction in 2005–6, when the DBT introduced a new scheme called the Small Business Innovation Research Initiative (SBIRI). The stated aim of the SBIRI is to provide financial assistance to biotech companies for R&D, techno-logical innovation, and commercialization.[76] Funding as per the SBIRI scheme operates in two phases. In the first phase, the SBIRI supports early stage, pre-proof-of-concept research. Funding options available to biotech companies in this phase depend on the total project cost, and the amount the company is able/willing to invest. In the second

[72] Loeppky, 'History, Technology, and the Capitalist State'; Bains, *Venture Capital*.

[73] Salter et al., 'Stem Cell Science in India', 84.

[74] GOI, 'Report of the Working Group for the Eleventh Five Year Plan', 72.

[75] Government of India, 'National Biotechnology Development Strategy, Key Elements: The Future Bioeconomy: Translating Life Sciences Knowledge into Socially Relevant, Eco Friendly and Competitive Products' (New Delhi, 2007), 1.

[76] GOI, 'Report of the Working Group for the Eleventh Five Year Plan', 62.

phase, the SBIRI provides financial support for technological innovation and commercialization.[77] Till 2008, the DBT had spent INR 570 million funding 48 projects under this scheme.[78] By 2011, according to the DBT, 91 projects were sanctioned under the SBIRI with an outcome of six Indian patents and development of 16 technologies in the field of agriculture, health, and instrumentation.[79]

The SBIRI scheme defines 'Small Business Unit' 'as an enterprise with not more than 500 employees in R&D'.[80] This is a rather generous definition of small enterprises. The DBT's motive behind this new definition of small enterprises is to bring even big biotech companies under the SBIRI scheme. In fact, many big biotech companies such as Nuziveedu Seeds, Bharat Biotech, Bharat Serum, Avesthagen, Cadila, and Reliance Life Sciences have obtained funding for their projects and have taken up projects under this scheme. Here too, Indian policy makers have taken their cue from their counterparts in the US. The US Bayh–Dole Act contained a provision that seemed to favour small companies. Inasmuch as this provision encouraged universities to identify small companies as the licensees of their patents on public-funded research outcomes, the Act could guise itself as a benefactor of small companies. Soon, however, this provision 'was quietly extended to *large* corporations by means of a 1983 executive memorandum by President Reagan'.[81] Interestingly, the DBT has defined small enterprises in such a manner that it also includes big biotech companies.

In the year 2008, the DBT officially extended its financial assistance programme to big biotech companies, when it introduced a new initiative

[77] In the first phase, the DBT provides 80 per cent grant for projects costing up to INR 2.5 million. If project cost is more than INR 2.5 million and less than INR 10 million, the DBT provides 50 per cent grants. When project cost goes beyond INR 10 million, the DBT provides INR 5 million grant as well as interest free loan to biotech companies. In the second phase, the DBT provides soft loan up to INR 100 million on a miniscule simple interest rate of 2 per cent. Details are sourced from http://www.sbiri.nic.in/HTML/funding_structure.html (accessed December 2013)

[78] Banerjee, 'India: Science & Technology 2008', 62.

[79] GOI, 'Report of the Working Group for the Twelfth Five Year Plan', 21.

[80] Available at http://dbtindia.nic.in/uniquepage.asp?id_pk=136 (accessed August 2013).

[81] Mirowski and Horn, 'Contract Research Organization', 522.

called the BT Industry Partnership Programme (BIPP). The stated goal of the BIPP was to make Indian biotech industry globally competitive by promoting path-breaking research in frontier, futuristic technology areas with enormous economic potential.[82] The original plan of the DBT was to let the BIPP function in a pilot mode for two years with a budget outlay of INR 3,500 million.[83] Under the BIPP scheme, the DBT provides 35–50 per cent financial assistance to biotech companies. Not surprisingly, the IP created by this programme would be owned by the biotech companies while the DBT would receive a small royalty of 5 per cent out of it.[84] According to the DBT, by the year 2011, 60 projects were financially supported under the BIPP which benefited 51 biotech companies and resulted in the development of H1N1 vaccine, production of a cost effective VLP-1 HPV vaccine, development of 'Herbicide & Stress tolerant' transgenic onion, and so on.[85]

The SBIRI and the BIPP schemes provide financial assistance, in the form of grants and interest free loans, to biotech companies for technological innovations. Any IP emerging out of these schemes belongs to biotech companies. The ownership of IP gives monopoly rights to corporations allowing them to fix prices of their products and services at exorbitantly high levels. This will exclude the vast majority of Indian population from accessing the goods and services produced by BT industry.

★ ★ ★

Neoliberal economic reforms in India have led to an increase in the wealth of a few business houses, real estate developers, rural landed elite, and benefited a section of the English educated urban middle-class. Rising GDP figures, stock-exchange booms and a high growth rate of the national economy provide a measure of these changes. This growth rate, however, masks state failure to address issues of inequality,

[82] GOI, 'Report of the Working Group for the Twelfth Five Year Plan', 107.

[83] Banerjee, 'India: Science & Technology 2008', 107.

[84] Details are available at http://dbtindia.nic.in/uniquepage.asp?id_pk=680 (accessed August 2013).

[85] GOI, 'Report of the Working Group for the Twelfth Five Year Plan', 21, 107–8.

poverty, unemployment, malnutrition, health, and education.[86] The inability of a large number of people to meet their basic nutritional requirements has led scholars such as Utsa Patnaik to term India a 'republic of hunger', with over three-fourths of the population living below the poverty line.[87] Neoliberal economic policies have further exacerbated the situation for those who depend on agriculture. Between 1980 and 2007, the agricultural sector's contribution to the GDP declined from 39 per cent to 17 per cent, whereas the number of people dependent on it only declined from 68 per cent to 57 per cent of the population.[88] This has in turn effected large-scale migration to cities. Statistics offered by the 2011 Census show that for the first time since 1921, urban India added more people to the total population than rural India.[89] Cities, on the other hand, have few jobs to offer to these rural migrants and a large number of them swell the bracket of the urban 'surplus people' living in slums and pavements.

In the last two decades, neoliberal reforms have also overdetermined the character of BT sector in India. The same economic, political, and ideological processes which are responsible for the growing inequality and poverty also underlie India's promotion of the biotech industry. Thus, the claim of Indian planners and policy makers that BT shall address issues of food, nutrition, and health of the vast majority of poor people sounds rhetorical. Throughout this chapter, we can in fact identify at least two basic senses in which the poor/poverty emerge within India's BT-related policy discourse. In the first, the poor appear as the ultimate/intended beneficiaries of BT. This is achieved by replicating a near universal and compulsory rhetoric of policies ushering in welfare of the poor (almost all of India's policies begin with such an invocation). This benefit-to-the-poor rhetoric is important for the legitimacy of not only BT but also most technologies. In the second,

[86] Amit Bhaduri, *Essays in the Reconstruction of Political Economy* (New Delhi: Aakar Books, 2010); P. Patnaik, *Re-envisioning Socialism*.

[87] Utsa Patnaik, *Republic of Hunger and Other Essays* (Gurgaon: Three Essays Collective, 2007).

[88] Amit Basole and Deepankar Basu, 'Relations of Production and Modes of Surplus Extraction in India: Part I-Agriculture', *Economic and Political Weekly* 46, no. 14 (2011): 41–58.

[89] Palagummi Sainath, 'Census Findings Point to Decade of Rural Distress', *The Hindu*, Delhi, 26 September 2011.

we see the Indian state blatantly advertising poverty as one of the most basic reasons for the global BT industry to consider India as a viable ground in which to relocate certain kinds of operations or aspects of production. The availability of relatively cheap skilled workforce and the existence of a vast number of disease-ridden poor from diverse genetic pools here becomes a reason for turning India into a hub for contract research and clinical trials. The poor/poverty in India thus appear both as reason for India's expected rise in the global BT sector and as the beneficiaries of the new technology.

9 Stem Cell Research and Experimentation in India

Leveraging Hope for Global Prominence

Rohini Kandhari

This chapter describes how state policies in India have enabled and legitimized clinical experimentation of stem cells whose therapeutic potential has not yet been fully realized or even understood. Through interviews with clinicians experimenting with adult stem cells, the chapter argues that the state, by fostering the promise of future cures from the techno-science, is drawing India's doctors and patients into the highly experimental fold of stem cell research. The argument presented here does not reject the potential of stem cells to relieve human suffering someday and neither does it deny the necessity of using human subjects for medical experiments. What it questions is if the state's scientific and economic aspirations are widening the nation's potential as a site for medical experimentation while overriding the health concerns of its people.

In the past decade, India has been attempting to build a legitimate, globally accepted, techno-scientific enterprise of stem cells. The state's concerted policy and legislative changes from the 1980s onwards, particularly in the related areas of health care, clinical trials, and intellectual property have enabled the growth of a biotechnology industry in which stem cells have a part to play. A recent report titled 'Indian

Biotechnology: The Roadmap to the Next Decade and Beyond', pre-pared by the Association for Biotech Led Enterprises (ABLE) for the Department of Biotechnology (DBT), includes stem cells in its 'frontier areas' of biotechnology.[1] The ABLE report suggests that India must seize the opportunity presented by emerging biotechnologies such as stem cells in order to build the nation's bio-economy and to realize its potential as a global innovation hub.[2]

The potential of stem cells to provide groundbreaking treatments for incurable conditions has captured the attention of actors as diverse as scientists, politicians, patients, biotech firms, and the media, the world over. The conditions that await cures from stem cells include Alzheimer's and Parkinson's disease, spinal cord injuries, diabetes, and heart conditions. Stem cells are unspecialized cells that are highly prized because of their potential to become some or any cell in the human body and for their ability to self-renew. This power of self-renewal gives stem cells the capacity to repair or regenerate damaged tissue. The most precious kind of stem cell is that found in the embryo. This is for a number of reasons. Embryonic stem cells are more likely to become any one of the tissue types in the human body. These cells also have the potential to renew themselves indefinitely, in culture or in vitro, over many years, without losing their undifferentiated quality. A collection of such healthy, undividing cells in a laboratory is a stem cell line that has the potential to provide an unlimited supply of cells, and ultimately tissue. As a result, many hope that stem cells, particularly embryonic stem cells, will someday also eliminate the need for organ donation.[3] The extraction of embryonic stem cells, however, requires the embryo to be destroyed in the early stages of its formation. 'Pro-life' supporters, particularly in the United States (US), oppose the deliberate destruction of the embryo on the grounds that the embryo is a form of human

---

[1] Prakash Satya Dash, *Indian Biotechnology: The Road Map to the Next Decade and Beyond* (New Delhi: Department of Biotechnology and Association of Biotechnology Led Enterprises, 2012).

[2] Dash, *Indian Biotechnology*, 7. ·

[3] Catherine Waldby, 'Stem Cells, Tissue Cultures and the Production of Biovalue', *Health* 6, no. 3 (2002): 305–23; Sarah Sexton, 'The Future Is Now: Genetic Promises and Speculative Finance', in *Global Health Watch 3: An Alternative World Health Report* (London: Zed Books, 2011), 199–210.

life.[4] In response to the moral debate around embryo destruction, the restrictive regulatory regime on human embryonic research that emerged in many parts of the world has brought to focus research on other types of stem cells.[5] These are adult stem cells found in the bone marrow and other tissue. Stem cells are also found in umbilical cord blood and the foetus. Each of these categories of stem cells differ in their ability to specialize into various cell types.[6]

Today, established treatments in stem cells are mainly confined to stem cells found in the blood, which are essentially bone marrow transplants for blood related disorders.[7] Umbilical cord blood transplants are still largely considered experimental procedures with only 2,500 transplants having been conducted the world over by mid-2003.[8] Research on adult stem cells (apart from bone marrow for accepted conditions) is in its early stages and so far there has been no successful Phase I[9] clinical trial using human embryonic stem cells.[10] Many researchers

---

[4] Melinda Cooper, *Life as Surplus: Biotechnology & Capitalism in the Neoliberal Era* (Washington: University of Washington Press, 2008).

[5] Divided views on embryonic stem cell research have resulted in a complex global regulatory regime that ranges from a complete ban on embryonic research in Ireland, to allowing the creation of embryos for research in countries like the United Kingdom (UK) and India—Herbert Gottweis, Brian Salter, and Catherine Waldby, *The Global Politics of Human Embryonic Stem Cell Science: Regenerative Medicine in Transition* (London and New York: Palgrave and Macmillan, 2009), 5. According to the Biophoenix 2006 report, only 11 of the 106 stem cell companies that were identified across the world prioritized human embryonic research—Biophoenix, 2006, cited in Gottweis, Salter, and Waldby, *Global Politics of Human Embryonic Stem Cell Science*, 27.

[6] Catherine Waldby and Robert Mitchell, *Tissue Economies: Blood, Organs, and Cell Lines in Late Capitalism* (Durham and London: Duke University Press, 2006).

[7] Australian Stem Cell Centre, 'Stem Cell Therapies: Now and in the Future', *Stem Cell Foundation*, last modified 7 December 2011, http://www.stemcellfoundation.net.au.

[8] Waldby and Mitchell, *Tissue Economies*.

[9] A Phase I trial tests a new intervention or treatment in small numbers of people (usually healthy volunteers) for the first time. The trial is conducted with the primary objective of evaluating the intervention's safety, its possible side effects, and to determine an acceptable dose range.

[10] Gottweis, Salter, and Waldby, *Global Politics of Human Embryonic Stem Cell Science*.

also point to the worrying cancerous properties of stem cells and their potential for inappropriate tissue formation, such as muscle developing in the brain, once stem cells are inserted into the body.[11]

Currently, therefore, the lure of stem cells lies in 'what they could do rather than what they are'[12] and stem cell research is 'justified by the promise of finding miraculous cures for debilitating diseases.'[13] This chapter argues that by fostering the promise—therapeutic, economic, and scientific—of stem cells, the Indian state is drawing India's doctors and patients into the highly experimental fold of stem cell research. It describes how state policies have enabled and legitimized clinical experimentation of both embryonic and adult stem cells whose therapeutic potential has not yet been fully realized or even understood. The argument presented here does not reject the potential of stem cells to relieve human suffering someday and neither does it deny the necessity of using human subjects for medical experiments. What it questions is if the state's scientific and economic ambitions are widening the nation's potential as a site for medical experimentation while overriding the health concerns of its people.

The chapter is divided into two parts. The first section discusses the state's role in facilitating the growth of India's biotechnology industry, with a focus on stem cell research in particular. The chapter describes how India's aspiration for global prominence in science is the reason why the state in alliance with the industry is promoting biotechnologies such as stem cells that are sustained by the 'continuous reproduction of the promise'.[14] The biotechnology sector is stated to become an important driver for the growth of India's economy. India's projected gross domestic product (GDP) of 41 trillion dollars by 2050 is expected to position the country in the top three world economies along with China and the US).[15] As envisioned by the Government of India's (GOI's) Department of Biotechnology, the biotech sector is expected to boost India's economy and simultaneously provide solutions to the country's

---

[11] Waldby, 'Stem Cells, Tissue Cultures', 317.

[12] Sexton, 'The Future Is Now', 4.

[13] Tiago Moreira and Paolo Palladino, 'Between Truth and Hope: On Parkinson's Disease, Neurotransplantation and the Production of the "Self"', *History of the Human Sciences* 18, no. 3 (2005): 67.

[14] Cooper, *Life as Surplus*.

[15] Dash, *Indian Biotechnology*.

food and health problems.[16] While India's developmental policies have historically emphasized the centrality of science and technology, the chapter argues that in the present, the state's need for science has moved far beyond its role in the socioeconomic and cultural upliftment of Indians. Today, the difference lies in the state leveraging its scientific strength that includes its human resources for advancing its global positioning in science, innovation, and the world economy.

The first part of this chapter foregrounds the second section that provides an insight into the practice of stem cell experimentation in India. The second section indicates how the key actors of the industry, the doctor and patient, are brought together in the context of hope for therapeutic breakthroughs in stem cell treatments. From primary research on clinicians in Delhi and Bangalore using stem cells, it seems that there are several doctors experimenting with adult stem cells. Experimentation here is understood as the unregulated provision of adult stem cell treatments for conditions such as cerebral palsy for which stem cells of any type have not yet proven to be safe or effective. It is also understood in the context of regulated clinical trials using adult stem cells.

INDIA: A 'GLOBAL SCIENTIFIC POWER BY 2020'

At the Indian Science Congress held in January 2012, India's Prime Minister, Manmohan Singh, expressed concern about 'India's relative position in the world of science' that 'has been declining' and being 'overtaken by countries like China'. He committed to increasing India's research and development (R&D) spending in science to at least two per cent of the GDP by the end of the Twelfth Five Year Plan or from the period between 2012 and 2017. This expenditure would only be possible if the industry increased its spending on R&D further, stated the Prime Minister.[17] India's current spending on science and

---

[16] Government of India, Department of Biotechnology, 'National Biotechnology Development Strategy', http://www.dbtindia.nic.in (accessed 19 August 2012).

[17] India Today Online, 'Full Text of PM's Speech at Indian Science Congress', *India Today*, 3 January 2012, http://indiatoday.intoday.in/story/mannmohan-singh-china-overtaken-india-in-field-of-science/1/166999.html (accessed 26 March 2012).

technology, constituting 1 per cent of its GDP is lower in comparison to China's expenditure that increased from 0.6 per cent of its GDP in 1995 to 1.2 per cent in 2004.[18] This funding structure for R&D stands in contrast to the US, where federal funding amounts to only about one-quarter of total R&D funding, indicating a greater presence of the private sector in the nation's R&D expenditure.[19] At the Science Congress, the Prime Minister also added that scientific research should be directed towards providing solutions to India's chronic problems of food, water, and electricity.[20]

Science and technology have historically been privileged by the Indian state for alleviating social ills and nation building. In the immediate aftermath of independence, Jawaharlal Nehru, India's first Prime Minister, stated that 'science alone ... can solve the problems of hunger and poverty, of insanitation and illiteracy'.[21] For the government of a newly independent India, science was the key transformative agent in a society they thought to be plagued by backwardness and superstition.[22] The manifesto of the Indian National Congress in 1945 proclaimed scientific research to be a 'basic and essential activity of the state' that 'should be organized ... on the widest scale'.[23] In the present day, the role of science and technology extends beyond addressing local or national needs. The nation's aspiration to become a leader in the scientific world has changed the way in which the state projects itself and shaped the methods by which it attempts to realize its goals.

---

[18] Organization for Economic Co-operation and Development, 'China Will Become World's Second Highest Investor in R&D by End of 2006, Finds OECD', http://www.oecd.org (accessed 4 December 2006).

[19] Paul Wilson and Aarthi Rao, 'India's Role in Global Health R & D' (Washington DC: The Results for Development Institute, 2012).

[20] India Today Online, 'Full Text of PM's Speech at Indian Science Congress'.

[21] Venni V. Krishna, 'A Portrait of the Scientific Community in India: Historical Growth and Contemporary Problems', in *Scientific Communities in the Developing World*, ed. Jacques Gaillard, Vipin Krishna and Roland Waast (New Delhi: Sage Publications, 1997), 236–80.

[22] Srirupa Roy, *Beyond Belief: India and the Politics of Postcolonial Nationalism* (Durham and London: Duke University Press, 2007), 109.

[23] Krishna, 'A Portrait of the Scientific Community in India', 237.

Kapil Sibal, in his tenure (2004–9) as Minister for Science & Technology stated:

> India is today perceived as a country that has the potential of becoming a global player in the biotechnology area. In order to achieve this, a major thrust is necessary for promoting innovation and commercialization. We have natural strengths to help us to take this sector forward, such as a large pool of manpower, well-developed scientific infrastructure … a large diverse heterogeneous human population and above all a well regarded brand value in the knowledge sector.[24]

The idea of India as a globally renowned centre of scientific expertise has formed the framework within which the nation's biotechnology policies have been strategized and implemented in the recent decade. In this envisioning of India as a powerhouse of scientific innovation and clinical research, Indians have become marketable assets rather than beneficiaries of state policy. In the course of this chapter it will be made evident that India's large patient population in addition to a highly skilled workforce are being leveraged by the Indian state as a source of strength that will help take the nation forward in biotechnology.

In 1986, India established a dedicated Department of Biotechnology, which, since its inception has been a major player in the nation's biotechnology industry. With an objective of making India 'more competitive in the life sciences' by promoting 'innovation and entrepreneurship' the DBT is today, the largest central government funding agency for the life sciences.[25] In 2011, the biotechnology sector was valued at US$4 billion, a six-fold increase since 2003.[26] With a more innovation friendly environment 'spurred' on by favourable government policies, India's biotechnology industry hopes to contribute 100 billion dollars to India's GDP by 2025.[27] While the country's R&D expenditure had always prioritized traditional funding areas such as defense and atomic energy, the DBT's budget over the years is indicative

---

[24] Government of India, Department of Biotechnology, 'Biotech News', last modified 2006, http://biotechnews.gov.in/readeditor.11t.html.

[25] Government of India, Department of Biotechnology, 'Annual Report', Department of Biotechnology, Ministry of Science and Technology, 2009–10, http://www.dbtindia.nic.in (accessed 19 August 2012), 1.

[26] Dash, *Indian Biotechnology*.

[27] Dash, *Indian Biotechnology*, 21.

of the state's simultaneous and increasing interest in biotechnology. At the time of the DBT's establishment, its budget was US$96 million rising to about US$358 million in 2004–5.[28] Within life sciences funding, state support for health biotechnologies has been gaining ground in recent years. The DBT's funds for medical biotechnology increased in the Tenth Five Year Plan period (2002–7) by 36 per cent of the total budget from 13 per cent in the Ninth Five Year Plan (1998–2002).[29]

In the 1990s, the health care biotechnology industry in India witnessed an expansion of private companies that adopted different processes for their establishment and growth. Some were created when the parent pharmaceutical company diversified into the biotechnology sector by establishing a biotech segment. Others were set up as dedicated health biotechnology start-up firms.[30] The entire sector today has 300 firms involved with agri-biotech, medical technologies, and bio-services or contract research and clinical trials. The biopharmaceutical and health care segment that includes vaccines, medical devices, and emerging technologies such as stem cells accounts for 60 per cent of the market share.[31]

The DBT's National Biotechnology Development Strategy lists stem cell engineering and regenerative medicine as one of the medical biotechnologies 'vital to India's progress'.[32] In 2005, the DBT together with the Indian Council of Medical Research (ICMR) launched a

[28] Sachin Chaturvedi, 'Dynamics of Biotechnology Research and Industry in India: Statistics, Perspectives and Key Policy Issues' (OECD Science, Technology and Industry Working Papers 2005/06, Paris, 31 May 2005), 21.

[29] Brian Salter, Melinda Cooper, Amanda Dickins, and Valentina Cardo, 'Stem Cell Science in India: Emerging Economies and the Politics of Globalization', *Regenerative Medicine* 2, no. 1 (2007): 83.

[30] Parveen Arora, 'Healthcare Biotechnology Firms in India: Evolution, Structure and Growth', *Current Science* 89, no. 3 (2005): 458–63; Sarah E. Frew, Rahim Rezaie, Stephen M. Sammut, Monali Ray, Abdallah S. Daar, and Peter A. Singer, 'India's Health Biotech at a Crossroads', *Nature Biotechnology* 25, no. 4 (2007): 403–17.

[31] Dash, *Indian Biotechnology*.

[32] Government of India, Department of Biotechnology, 'Vision', Department of Biotechnology, Ministry of Science and Technology, available at http://www.dbtindia.nic.in/about-us/vision-and-strategy (accessed 19 August 2012).

national stem cell initiative. In the same year, the Department formed a National Task force on stem cells. This body was responsible for carrying out the goals of the national stem cell initiative that included promoting therapeutic applications of stem cells and developing stem cell city clusters. Currently, the DBT's plan for the stem cell industry includes supporting basic research and promoting research with the objective of developing clinical applications of both embryonic and adult stem cells. Other focus areas include investing in centres of excellence and establishing autonomous institutions for stem cell research. By 2007, the DBT had supported more than 30 stem cell research projects that included the use of both embryonic and adult stem cells.[33]

In India today, there are over 40 institutions in the country, across both public and private sectors that are involved in stem cell research.[34] Their practices vary from basic research to clinical trials and stem cell treatment using adult and embryonic stem cells. Some public institutions like the National Centre for the Biological Sciences (NCBS) have gained global repute for creating human embryonic stem cell lines. In 2001, the US President, George Bush announced that federal funds for human embryonic stem cell lines must be restricted to those cell lines that were already in existence. Globally, only 64 cell lines met this criterion. Of this figure, ten cell lines were found in India of which three belonged to the NCBS.[35] The Jawaharlal Nehru Centre for Advanced Scientific Research in Bangalore has also created two human embryonic stem cells lines that have been deposited in the stem cell bank in the UK. This bank functions as a repository for quality controlled stem cell lines that are available to the global scientific community.[36] Another public institute in Bangalore conducting both basic and translational stem cell research is the Institute for Stem Cell Biology and Regenerative Medicine or inStem. InStem is among seven

[33] Salter et al., 'Stem Cell Science in India', 82.

[34] Dash, *Indian Biotechnology*, 34.

[35] Aditya Bharadwaj and Peter Glasner, *Local Cells, Global Science: The Rise of Embryonic Stem Cell Research in India* (London: Routledge, 2009), 20.

[36] Jawaharlal Nehru Centre for Advanced Scientific Research, 'JNCASR Scientists Derive Human Embryonic Stem Cells from Discarded Embryos', posted 25 August 2008, http://www.jncasr.ac.in/newsview.php?id=51 (accessed 22 September 2012).

new autonomous institutions of the DBT that were established during the Eleventh Five Year Plan period (2007–12).

Together with NCBS, inStem constitutes the Bangalore Bio-cluster and represents DBT's attempt at innovative models of partnership and governance. Bio-clusters are being promoted by the DBT in order to create research environments that enhance cost-effective translational projects and collaborative efforts in biotechnology.[37] In addition to inStem's own research objectives, the institute is meant to function as an umbrella body for an extramural funding programme intended to support stem cell research across the country and the Center for Stem Cell Research at the Christian Medical College in Vellore.[38] In addition to these public institutions conducting stem cell research, is the All India Institute of Medical Sciences (AIIMS) in New Delhi. A tertiary, teaching hospital, AIIMS conducts both basic and translational stem cell research. In 2005, the institution established its stem cell facility that is supported by the DBT.[39]

Among the private enterprises that have invested in stem cell research is Reliance Life Sciences in Mumbai. The company holds the remaining seven of the ten embryonic stem cell lines identified in India after President Bush announced his stem cell regulations.[40] Another private stem cell company is Bangalore's Stempeutics Research Pvt. Ltd. Its website describes the firm as a 'leading stem cell company developing stem cell based medicinal products' with a mission 'to transform medicine and offer new hope to millions of people'. In 2009, Stempeutics conducted a combined Phase I & II[41] stem cell clinical trial in India

---

[37] Government of India, Department of Biotechnology, 'Report of the Working Group for the Eleventh Five Year Plan (2007–2012) (New Delhi, 2006), www.dst.gov.in/about_us/11th-Plan (accessed 14 August 2013).

[38] InStem (2010), http://www.instem.res.in (accessed 21 August 2012).

[39] All India Institute of Medical Sciences (AIIMS), 'Introduction', *All India Institute of Medical Sciences*, http://www.aiims.edu/aiims/stemcell/introduction.htm (accessed 6 January 2013).

[40] Bharadwaj and Glasner, *Local Cells, Global Science*.

[41] Combined Phase I and II trials are done on populations for whom all therapeutic options have been exhausted. Patients suffering from diseases such as cancer or HIV/AIDS are selected for the trials instead of healthy volunteers. This is because the toxicity of anti-retroviral drugs or anti-cancer drugs could expose healthy volunteers to unnecessary risks that would outweigh

to test its stem cell product for Critical Limb Ischemia or for patients with chronic non-healing wounds. The stem cells used were those donated from bone marrow. Stempeutics has collaborated with the Indian pharmaceutical company CIPLA to market this product.[42]

In 2010, India's stem cell market was worth US$450 million and is currently growing at the rate of 15 per cent.[43] Although India's presence in the US$70 billion global biotechnology industry is still negligble, its stem cell industry is perceived as an important contributor to developing India's bio-economy.[44,45] Stem cell research constitutes a part of the scientific innovation that has been taking place in the knowledge economies of the developed world for the past two decades. In the 1980s, when countries like the US were confronted with stagflation, their business communities began to explore the commercial potential of research emerging from the life sciences and informatics.[46] New biotechnologies such as stem cell science that are highly capital and knowledge intensive required considerable state support for building networks among diverse interests such as private and public research funding, science education, venture capital, intellectual property, and bioethics. Using a 'pragmatic mix of neoliberalism and coordinated approaches', states of the OECD[47] actively created conditions for the commercialization of life science research.[48] Stem cell research and development is illustrative of this multiple institutional and tactical support provided by governments the world over to capitalize on the

---

the benefits of the research. Government of India, Indian Council of Medical Research, *Ethical Guidelines for Biomedical Research on Human Participants* (New Delhi: Indian Council of Medical Research, 2006).

[42] Stempeutics Research Private Ltd, last modified 16 March 2012, http://www.stempeutics.com (accessed 21 August 2012).

[43] Dash, *Indian Biotechnology*, 34.

[44] Dash, *Indian Biotechnology*, 36.

[45] Ernst & Young, 'Beyond Borders: Global Biotechnology Report 2007' (UK: Ernst & Young, 2007).

[46] Cooper, *Life as Surplus*, 15–50.

[47] Organisation for Economic Co-operation and Development is an international economic organization with a membership of 34 countries.

[48] Gottweis, Salter, and Waldby, *Global Politics of Human Embryonic Stem Cell Science*, 29.

commercial prospects of the techno-science. Many deliberate policy and legislative changes were thus initiated at this time by developed nations to gain competitive advantage in the market potential of new discoveries in human biological processes.[49]

While India still 'has a long way to go' in the emerging area of regenerative medicine and tissue engineering, the country does have 'the potential to emerge as a global leader' in the field, stated the ABLE report.[50] India's potential for securing leadership positions in different areas of science, technology, and innovation is articulated in several state policy documents. For example, in 2010, the Science Advisory Council to the Prime Minister Manmohan Singh released its visionary document, 'India as a Global Leader in Science'. Indian science has many challenges to overcome, stated the Advisory Council, 'yet, there are good reasons why 'India's presence in the world of science cannot be ignored'.[51] One of these reasons is the country's large young population. Across science policy documents, India's youth is touted as a major advantage for leveraging India's imminent presence in the global scientific community. The key objective of the 'Science, Technology and Innovation (STI) Policy 2013' is 'positioning India among the top five global scientific powers by 2020'.[52] The policy describes India's 'large demographic dividend' and 'huge talent pool' as an important asset in building the country's innovative capacity and scientific future.[53]

Science education is thus a major focus area of India's biotechnology policy. The DBT supports science education at the school and university level and also awards scholarships and cash prizes to biotechnology students and young scientists. India has the largest number

[49] Gottweis, Salter, and Waldby, *Global Politics of Human Embryonic Stem Cell Science*, 1–34.

[50] Dash, *Indian Biotechnology*, 36.

[51] Government of India, Department of Science and Technology, 'India as a Global Leader in Science' (New Delhi: Department of Science and Technology, 2010), 3.

[52] Government of India, Ministry of Science and Technology, 'Science, Technology and Innovation Policy 2013', Policy Document (New Delhi: Ministry of Science and Technology, 2013), 4.

[53] Government of India, Ministry of Science and Technology, 'Science, Technology and Innovation Policy 2013', 1.

of physical/biological PhD students in absolute terms compared to its Asian competitors like China and South Korea. China, however, seems to have superseded India in terms of the rate of increase in the production of PhD degrees.[54] Moreover, several post graduate students in the life sciences leave India every year to purse PhDs abroad. Among the reasons cited for this annual brain drain are inadequate training opportunities in India's academic institutions. Many students do not return to India once they have received their degrees. The DBT has made concerted efforts to encourage the expat scientist community to return to their home country.

For example, the Ramalingaswami Fellowship initiated in 2008 is a 're-entry grant' for scientists of Indian origin interested in research positions in India's biotechnology industry.[55] In the same year, the DBT also announced a collaboration with the UK-based Wellcome Trust—a five year, US$140 million grant for scientists. The grant offered attractive annual salary packages ranging from US$16,000 to US$30,000 a year for a period of three to five years, with the choice of working at any Indian institution.[56] In an attempt to retain fresh graduate and postgraduate biotechnology students in India, the DBT's Biotech Industrial Training Programme gives young scientists hands-on training opportunities in the industry. Held in collaboration with the Biotech Consortium of India Limited (BCIL) the project requires students to undergo a six-month training period after which they find placements in biotech firms who also benefit from easy access to trained manpower.[57]

Promoting closer ties with the industry to nurture innovation and accelerate the commercialization of research is among DBT's key strategies. Soon after the Department's establishment, the BCIL was formed as a public limited company in order to accelerate the building of a relationship between research institutes and the industry through technology transfers.[58] By 2005, the BCIL had facilitated 60 technology

---

[54] Salter et al., 'Stem Cell Science in India', 75–89.

[55] Government of India, Department of Biotechnology, 'Annual Report'.

[56] Yudhijit Bhattacharjee, 'India Hopes New Fellowships Will Attract Expat Scientists', *Science* 312, no. 5895 (2008): 1431.

[57] Government of India, Department of Biotechnology, 'Annual Report'.

[58] Salter et al., 'Stem Cell Science in India', 86.

transfers between the public and private sectors.[59] Additionally, many initiatives have been undertaken to support biotechnology innovation and enterprise. For example, the DBT's Entrepreneurship Development Programme, implemented across eight states, trains entrepreneurs in business management skills to establish commercial biotech enterprises. The DBT also awards cash prizes for product and process development to individual scientists and companies both in the public and private sectors.

In 2007, the Government of India increased its funding to public–private partnerships for biotech and pharmaceutical R&D, allocating 30 per cent of the DBT's budget to public–private agreements.[60] The DBT's National Biotechnology Development Strategy that was also announced in 2007 stated the need to support innovation in small and medium biotech enterprises. The strategy proposed the establishment of a Biotechnology Industry Research Assistance Council that would be responsible for promoting biotech start-ups and ensuring the conversion of public-funded and privately generated research sectors into 'viable and competitive enterprises'.[61] Due to the negligible presence of venture capital funding for Indian biotech the BCIL that was set up for advancing the commercialization of publicly funded research was also meant to function like a venture capital firm—providing venture capital to new biotech entrepreneurs.[62]

Unlike the US where venture capitalists are a dominant feature in biotechnology, in India, their presence in the sector is limited and steps are being taken to bring venture capital to the industry. For example, the State of Karnataka's Millennium Biotech Policy II announced a

---

[59] Priya Ranjan, 'The Political Economy of Technoscience: A Study of Medical Biotechnology in India', unpublished PhD thesis, Centre of Social Medicine and Community Health, Jawaharlal Nehru University, New Delhi, 2013, p. 120.

[60] Government of India, Department of Biotechnology, 'Biotechnology Industry Partnership Programme', Department of Biotechnology, Ministry of Science and Technology, 2008, available at http://www.dbtindia.nic.in (accessed 3 September 2008).

[61] Government of India, Department of Biotechnology, 'Biotechnology Industry Partnership Programme'.

[62] Salter et al., 'Stem Cell Science in India: Emerging Economies and the Politics of Globalization', *Regenerative Medicine* 2, no. 1 (2007): 75–89.

collaborative bio-venture fund of INR 50 crore (approximately US$8 million) with a private venture capital company to support 'hi-tech areas with strong social relevance' such as stem cell biology.[63] Another public–private funding scheme that supports cutting-edge technologies such as stem cells and tissue engineering is the DBT's Biotechnology Industry Partnership Programme (BIPP). The BIPP is a cost-sharing initiative in the form of loans and grants for high-risk innovation in biotechnology. Started in 2008, the goal of the BIPP is to make 'Indian industry globally competitive' and generate intellectual property in 'frontier, futuristic' technologies.[64] Although the government will bear the significant risk of the scheme contributing between 30 and 50 per cent to the grant-in-aid, the intellectual property rights will belong to the company. The publicly funded scientist who participates in this initiative will only receive a royalty on the patent.[65]

It is not surprising that the Indian state took active steps to promote the presence of the private sector in the country's biotechnology industry. The DBT was established at the time of India's initiation into the ideology of neoliberalism. Neoliberalism is described by Ralph Miliband as the process by which the relationship between state and private capital was consolidated with state intervention ensuring the continuing presence of capitalist enterprise.[66] As mentioned earlier, the global biotechnology industry was one such enterprise that developed as a result of calculated policy decisions made in the 1980s when neoliberalism was emerging as a major political force in many nations, particularly the US. Miliband argued that neoliberalism, contrary to popular belief, did not require the state to retreat in order to make way for the market and the two entities are not opposing forces.[67]

[63] Government of Karnataka, Department of Information Technology, Biotechnology and Science & Technology, 'The Millennium Biotech Policy II', *Department of IT, BT and S&T Karnataka*, http://www.bangaloreitbt.in/docs/2012/09/Biotech_Policy_II.pdf (accessed 23 September 2012).

[64] Government of India, Department of Biotechnology, 'Annual Report'.

[65] Government of India, Department of Biotechnology 'Biotechnology Industry Partnership Programme'.

[66] Ralph Miliband, *The State in Capitalist Society* (New Delhi: Aakar Books, 2011), ix.

[67] Government of India, Department of Biotechnology, 'Annual Report'; Miliband, *The State in Capitalist Society*, xiv.

Indeed, the large scale commercialization of public-funded scientific research in the US through its patent laws (explained in the following paragraphs) occurred as a result of state promotion of property rights in knowledge creation.

For India too, according to the vision document of the Science Advisory Council, enhancing patenting activity was important if it wanted to 'make an impact on' world science. According to the Science Advisory Council, India must increase its global patent filings to 20,000 patents per year by 2020 compared to the previous 1,900 patents filed by Indians in 2007.[68] India is behind both China and South Korea in terms of the number of patent filings within the country and outside. Regarding its ability to translate investment in R&D into patent filings, India also lags behind its Asian counterparts. In 2004, India's ranking in resident patent filings per million dollars of expenditure in R&D was thirtieth in the world compared to China that was ranked twelfth and South Korea first.[69] Patents are state granted, temporary monopolies given to patent holders on the 'commercialization of their research, allowing them to license use of their knowledge and materials for fees and loyalties'.[70] In other words patent protection means that no other company can sell or market the product for a fixed period of time.

Patents are important for investors in highly promissory bio-technologies such as stem cell research because intellectual property rights affords them some protection from the risks and uncertainties that underlie this techno-science. In the 1980s, significant reforms in America's patent laws transformed the global biotechnology and pharmaceutical industries. Among the legislative initiatives was the Bayh–Dole Act of 1980 that encouraged university–industry collabo-ration to ostensibly bring the fruits of research faster to the people. The act granted academic institutions the right to acquire patents on their research and then license them to companies. In the process, this led to rampant commercialization of research. In addition to legisla-tive change, the American judiciary also facilitated the privatization of public-funded research. In the landmark judgment of the *Diamond v. Chakrabarty* case in 1980, the US Supreme Court permitted a patent

[68] Government of India, Department of Biotechnology, 'Vision'.

[69] Salter et al., 'Stem Cell Science in India', 75–89.

[70] Gottweis et al., *Global Politics of Human Embryonic Stem Cell Science*.

on a genetically engineered microorganism. Since this historic ruling, intellectual property rights have been granted to embryonic stem cell lines turning biomedical research into an attractive investment opportunity.[71]

Since the passing of the Bayh–Dole Act, the number of patents granted to American universities had risen from less than 300 a year to more than 3,000 annually. These figures according to Bhaven Sampat 'downplay' other ways in which public-funded research can 'contribute to economic growth and innovation'.[72] These include the dissemination of ideas and transfer of technologies through peer reviewed publishing and teaching. In 2008, the Indian state produced its own version of the Bayh–Dole Act to facilitate the commercialization of public-funded research. The proposed bill that has not yet been passed by parliament has faced opposition due to its intention to make it mandatory for all academic institutions to patent the research they produce. Sampat warns developing countries wanting to accelerate the commercialization of academic research to be wary of 'aggressive patent policies' promoted by legislations such as the Bayh–Dole Act. He argues that patent laws must be based on local needs and developing countries should decide whether patents are a way of protecting domestic markets or of giving away to multinational companies the benefits of public-funded research.[73]

By the 1990s, India had committed to a stringent patent regime when it signed the Trade Related Intellectual Property Rights or TRIPS Agreement. The TRIPS agreement has been described as 'one of the most important neoliberal developments in global economic regulation ... representing the construction of public international law by, and in the exclusive interests of, a tiny handful of transnational corporations....'[74] Under the aegis of the World Trade Organization, TRIPS required a patent on all 'commercially exploitable products and

---

[71] Waldby and Mitchell, *Tissue Economies*.

[72] Bhaven N. Sampat, 'Lessons from Bayh–Dole', *Nature* 468, no. 7325 (2010): 756.

[73] Sampat, 'Lessons from Bayh–Dole'.

[74] David Tyfield, 'Neoliberalism, Intellectual Property and the Global Knowledge Economy', in *The Rise and Fall of Neoliberalism: The Collapse of an Economic Order*, ed. Kean Birch and Vlad Mykhnenko (London and New York: Zed Books, 2010), 60.

processes' for a minimum period of twenty years.[75] The Agreement was implemented with the intention of tightening patenting in the life sciences and instigating the global harmonization of minimum standards in intellectual property rights. These new patent rules imposed by TRIPS were justified by the global pharmaceutical industry as necessary for fostering innovation in knowledge-based sectors like biotechnology. It was also argued that a time-specific monopoly on marketing rights for drug inventors would help the industry recover huge R&D costs.

For India, however, signing TRIPS had serious implications for public health and threatened people's access to affordable medicines. Prior to signing TRIPS many developing countries like India did not require patents on medicines. India's patent act of 1970 enabled its indigenous drug industry to legally produce drugs patented abroad through a process called 'reverse-engineering'. These copies of patented drugs could then be sold cheaply to Indians. In 2005, when India had to amend its patent laws in compliance with TRIPS it meant increasing prices of generic drugs in the country because generic drug manufacturers would now have to pay royalties or a license fee to patent holders.[76]

In India's post-TRIPS environment, there have been several mergers of local drug companies with multinational firms and also acquisitions of Indian industry by foreign companies. India's recognition of product patents has given global drug companies some assurance of protection for their inventions. The generic industry has also been exploring different R&D strategies in order to maintain profits and competitiveness. These include in-licensing or out-licensing agreements and the contract services business model. An in-licensing agreement allows an Indian drug company to market a product for a multinational firm in India. It can also involve an Indian company producing a drug locally for a foreign company for a share of the profit. An out-licensing agreement occurs when Indian pharmaceutical companies license out their drug molecules to foreign firms to be developed further. Apart from these two kinds of collaborative arrangements, many Indian firms are focusing on providing contract research services. These involve conducting

[75] Tyfield, 'Neoliberalism, Intellectual Property', 64.

[76] Rajnish Kumar Rai, 'Battling with TRIPS: Emerging Firm Strategies of Indian Pharmaceutical Industry Post-TRIPS', *Journal of Intellectual Property Rights* 13, no. 4 (2008): 301–17.

and monitoring of clinical trials for the global pharmaceutical industry in India.[77] As a result of this realignment in India's drug industry, the clinical trial has become central to the current business framework as a new drug or interventional product cannot be developed without testing it on humans. 'In other words, the Indian pharmaceutical industry has itself served as a spur to the CRO[78] [contract research organization] sector.'[79]

The expansion of clinical trial operations in India was largely facilitated by an amendment, in 2005, to the rules of the Drugs and Cosmetics Act. The amendment allowed foreign drug companies to conduct clinical trials in the country concurrently with other trials being run outside India. Prior to the liberalization of the country's drug development laws, clinical trials for new drugs of foreign origin were permitted only if the later phase had already occurred in a foreign country. For example, Phase II trials—conducted to primarily establish efficacy of an intervention in patients—could be conducted in the country on the condition that the confirmatory phase or Phase III trials had taken place elsewhere. These regulations were important safeguards that prevented the country's population from being treated as first line trial subjects. The amendment removed the protective 'phase-lag' crucial for the safety of clinical trial subjects. It meant that now new drugs of foreign origin—those that have not yet been approved—could be conducted in India, while they were simultaneously also being tested elsewhere.[80]

The 2005 amendment to India's drug development laws significantly altered the nature and scope of the clinical trial in the country. While the traditional clinical trial regions of North America, Western Europe, and Oceania together comprise 66 per cent of the of the world's trial sites, countries such as China and India having grown rapidly in recent

[77] Rai, 'Battling with TRIPS'.

[78] Contract research organizations offer services to trial sponsors that include trial site management, data management, clinical trial design, medical writing, recruiting trial subjects, and investigators.

[79] Kaushik Sunder Rajan, 'Experimental Values, Indian Clinical Trials and Surplus Health', *New Left Review* 45 (May–June 2007): 67–88.

[80] Samiran Nundy and Chandra M. Gulhati, 'A New Colonialism? Conducting Clinical Trials in India', *The New England Journal of Medicine* 352, no. 16 (2005): 1633–6.

years.[81] The US Food and Drug Administration's figures from 2006 on the number of clinical trial investigators enrolled with the agency show that Russia had 623, the largest number of investigators, followed by India who had 464 investigators.[82] Today, the Indian bio-services sector that comprises CROs is the second largest sector after biopharmaceuticals in the biotechnology industry. It contributes 19 per cent of the total revenue generated by the industry.[83]

Transnational drug firms welcomed the changes in India's domestic policies regarding patents and drug development. Today, about 60 per cent of the global pharmaceutical market share belongs to the drug industry operating in the US that includes both European and American drug companies.[84] Faced with patent expiries of hugely profitable brand name drugs, lack of innovations in the drug pipeline and high infrastructure costs, these companies needed to expand their clinical research operations to low-cost countries such as India. The unavailability of large numbers of willing and suitable trial subjects in the developed world is also among the key reasons for the globalization of clinical trials. The 'treatment saturated' US population consuming many more drugs than it used to was rendered unsuitable to participate in clinical trials. The industry therefore needed 'treatment naïve' populations or those who are not taking drugs at the time of a clinical trial as interactions of different types of drugs can interfere with research outcomes. Recruiting large numbers of trial subjects is among the most costly and time-consuming aspects of running a clinical trial. When American prison populations were prohibited from participating in clinical trials, in the 1970s, US trial sponsors lost their largest human resource for clinical research.[85]

[81] Fabio A. Thiers, Anthony J. Sinskey and Ernst R. Berndt, 'Trends in the Globalization of Clinical Trials', *Nature Reviews Drug Discovery* 7, no. 1 (2008): 13–14.

[82] Adriana Petryna, *When Experiments Travel: Clinical Trials and the Global Search for Human Subjects* (Princeton: Princeton University Press, 2009), 13.

[83] Dash, *Indian Biotechnology*.

[84] Petryna, *When Experiments Travel*, 208.

[85] Sunder Rajan, 'Experimental Values, Indian Clinical Trials and Surplus Health', 67–88; Petryna, *When Experiments Travel*, 19–24; Charlotte Harrison, 'The Patent Cliff Steepens', *Nature Reviews Drug Discovery* 10, no. 1 (2011): 12–13.

India has been projected by both state and the industry as a favourable destination for clinical trials primarily because of its vast 'treatment naïve' population. The country's 'genetically diverse population of more than one billion people who have not been exposed to many medications but have myriad diseases, ranging from tropical infections to degenerative disorders' are perceived as ideal subjects for human experimentation.[86] This diverse patient population who appear to be suffering from a lack of adequate healthcare are therefore understood as an advantage for building India's bio-economy rather than a concern that must be addressed. Clinical trials are being conducted by physician–investigators across the entire spectrum of health care provision both in the public and private sectors. On 30 July 2013, 3,849 trials were registered with the Clinical Trial Registry of India (CTRI). Of this total number of trials, likely to be a conservative figure since registration is not a legal requirement, 60 were stem cell trials. The majority of stem cell trials registered are Phase I and II trials using adult stem cells.[87] A Phase I trial is the first step in the medical experimentation process with human beings. It is conducted to establish the basic safety of the intervention. In other words, Indians could be subjected to a particular stem cell being introduced into the human body for the first time. A Phase II trial evaluates the intervention's efficacy and further investigates its safety in human beings.[88]

The state promotion of cutting-edge biotechnology when majority of its population's health needs are unmet is a manifestation of a more aggressive embrace of market-based solutions for social sectors like health and education that took place in India in the 1990s.[89] Health sector reforms initiated in the country at this time led to shifting the responsibility of secondary and tertiary care to the private sector. At the time of India's neoliberal healthcare reforms, its health system was already privatized and highly iniquitous with urban–rural, caste, class, and gender disparities in terms of distribution of resources and health outcomes. The Indian government, recognizing the problems of health

[86]   Nundy and Gulhati, 'A New Colonialism?', 1634.

[87]   Government of India, National Institute of Medical Statistics, 'Clinical Trial Registry-India (2012)', last modified 8 May, http://www.ctri.nic.in/clinicaltrials/login.php (accessed 14 March 2013).

[88]   Government of India, Indian Council of Medical Research, *Ethical Guidelines.*

[89]   Bharadwaj and Glasner, *Local Cells, Global Science.*

inequity in the country had committed to providing comprehensive primary health care at the WHO Alma Ata declaration in 1978.

By the 1980s, however, India's attempt to provide universal health care irrespective of people's ability to pay was replaced by increased state support for the private sector and the simultaneous dismantling of the public sector. Due to the poor functioning of public health services, people were compelled to use the private sector and the health sector reforms in the 1990s only consolidated the process of privatization and commercialization of health care. Today, India's private sector has grown indiscriminately and without regulation resulting in a hetero-geneous mix of individual practitioners, clinics, nursing homes, and corporate hospitals that vary in size, cost, and in quality of care. Within this healthcare paradigm, health is viewed as a commodity that people must buy with the state's role being confined to providing minimal preventive services for the poorest sections that are unable to afford healthcare.[90] About 71 per cent of spending on health in India today is out of pocket and this expenditure pushes four per cent of India's population into poverty every year.[91] With health care expenses being one of the leading causes of poverty in India, for many patients, clinical trials hold the promise of treatment and perhaps even a cure.

In the past five years approximately 2,262 people have died in drug trials.[92] From media reports we know that trial subjects are largely India's poor and vulnerable populations. They are induced into trials by promises of free treatment, monetary incentives, and an implicit faith in their doctor's judgement.[93] In addition to clinical trials many patients

[90] Rama Baru, *Private Health Care in India: Social Characteristics and Trends* (New Delhi: Sage Publications, 1998), 40, 52; Imrana Qadeer, 2010, *New Reproductive Technologies and Healthcare in Neo-liberal India: Essays, Monograph* (New Delhi: Centre for Women's Development Studies), 11.

[91] Mohan Rao et al., 'Human Resources for Health in India', *The Lancet*, 377, no. 9765 (2011): 587–98.

[92] Press Trust of India, 'Post Stringent Norms, Clinical Trials in India Plummet', *Hindu*, Delhi, updated 22 April 2013, http://www.thehindu.com/sci-tech/health/policy-and-issues/post-stringent-norms-clinical-trials-in-india-plummet/article4639976.ece (accessed 23 July 2013).

[93] Jennifer Kahn, 'A Nation of Guinea Pigs', *Wired.com*, last modified March 2006, http://archive.wired.com/wired/archive/14.03/indiadrug.html (accessed 18 June 2011).

are also being exposed to highly risky, unregulated medical treatments in the private sector. With regard to stem cells, private practitioners across the country are providing stem cell treatments for several conditions that currently have no clinically proven stem cell applications. Several miraculous cures using stem cells have been claimed in the media and untested stem cell therapies are also advertised by Indian clinics via the Internet.[94]

In an attempt to regulate unethical practices by India's medical community and harmonize ethical regulations with western standards, the ICMR in 2006 published its ethical guidelines for biomedical research involving human subjects.[95] The following year, the DBT together with ICMR formulated ethical guidelines for stem cell research in 2007. These guidelines were updated in 2012 and finalized later in 2013.[96] The guidelines require an evaluation of stem cell research conducted in institutions by an Institutional Committee for Stem Cell Research and Therapy and a National Apex Committee for Stem Cell Research and Therapy to review restricted areas of research such as the use of human embryos for research. The fundamental ethical premise on which these guidelines are based is the trial subject's right to informed consent without undue inducement. Informed consent is the process by which potential trial subjects or donors of biological material for experiments are given the opportunity to exercise their free will with regard to participating in a clinical trial. 'It is through informed consent that the investigator and the subject enter into a relationship, defining mutual expectations and their limits....'[97] Informed consent is however,

---

[94] Noemie Bisserbe, 'A Shot in the Dark: Stem Cell Therapy Is Big Business in India. But Does the Hype Stem from Science or Faith?' *Business World*, 14 June 2010, http://www.businessworld.in/en/storypage (accessed 18 March 2013).

[95] Government of India, Department of Biotechnology and Indian Council of Medical Research, 'Guidelines for Stem Cell Research and Therapy' (New Delhi: Director General, Indian Council of Medical Research, 2007).

[96] GOI, Department of Biotechnology and ICMR, 'Guidelines for Stem Cell Research and Therapy'; Government of India, Indian Council of Medical Research and Department of Biotechnology, 'Guidelines for Stem Cell Research (Draft)', Draft Report (New Delhi: Indian Council of Medical Research, 2012).

[97] J. Robert Levine, *Ethics and Regulation of Clinical Research*. (New Haven and London: Yale University Press, 1988), 98.

not protection enough in an environment of medical experimentation where people are willing to take great risks in the hope for a cure.

The second part of the chapter looks at how doctors and their patients are brought together in a 'political economy of hope' or by the promissory value ascribed to the stem cell. The 'political economy of hope', a term used by Rose and Novas[98] and others is 'sustained through numerous reiterative practices' that involve a range of actors such as biotech companies, patients, clinicians, and policy makers who come together for the 'common cause' of keeping their options open for treatments in the future.[99] The Indian state's efforts to pursue good clinical practice in stem cell research only seek to legitimize such experimentation with stem cell treatments rather than securing the protection of the sick. Bharadwaj and Glasner argue that the state formulating ethical guidelines in accordance with international regulations are but strategic steps towards capturing global markets in stem cell research and ensuring investors of India's commitment to a regulated and reliable industry.[100]

EXPERIMENTING WITH HOPE: OPENING PANDORA'S BOX?

The hope for a cure and the expectation that better and improved stem cell treatments are 'always about to come' is integral to the survival and legitimacy of emerging biotechnologies such as stem cells.[101] The Indian state by nurturing this hope through calculated policy decisions has enabled and seemingly fuelled the clinician's desire to experiment with this little understood medical phenomenon. Moreover, it has raised the hope of patients who increasingly have to negotiate health care in a largely unregulated private sector.

[98] Nikolas Rose and Carlos Novas, 'Biological Citizenship', in *Global Assemblages: Technology, Politics, and Ethics as Anthropological Problems*, ed. A. Ong and Stephen J. Collier, 439–63 (Blackwell Publishing Ltd., 2005), 442.

[99] Alan Petersen and Kate Seear, 'Technologies of Hope: Techniques of the Online Advertising of Stem Cell Treatments', *New Genetics and Society* 30, no. 4 (2011): 332.

[100] Bharadwaj and Glasner, *Local Cells, Global Science*, 23–32.

[101] Nik Brown, 'Shifting Tenses: Reconnecting Regimes of Truth and Hope', *Configurations* 13, no. 3 (2005): 331–55.

'India has done a lot of work in stem cell research … but it is time that we take a leadership role in this. Please do not give the leadership role to the West that is my sincere request,' said a medical professional at the ninth annual conference of the Society for Regenerative Medicine and Tissue Engineering (SRMTE) held in Bangalore in 2013. The doctor believed that Indians 'are very capable people … but somehow' they cannot 'put together documents' or conduct clinical trials 'in a very disciplined way'. A vascular surgeon also present at the conference had similar thoughts. India 'can take a lead' he said in stem cell science because 'we have lot of patients, lot of data' and also a 'lot of need'. The problem however, the surgeon said, lies in India's 'terrible' record in documenting and in producing academic papers. He explained that India must increase its publications on stem cells because 'the world does not recognize us unless there's something in some peer reviewed journal. Which is also right to some extent. How do you prove your data?' The doctor had treated 'low-income' patients suffering from Buerger's disease with stem cells. The patients who were not charged for the treatment are believed to have experienced relief from pain and the healing of their wounds indicated some statistical significance. Follow up of these patients had, however, not been possible after six months of treatment. The experimentation took place without the availability of basic and requisite data from pre-clinical research. When questioned about the lack of even animal studies prior to his providing stem cell treatment, the surgeon was defiant. 'I don't care, my patient's doing well,' he said. He also admitted that it was 'not necessarily the right answer' and because experimenting with stem cells is 'so simple' 'all and sundry' are doing it and therein lies 'the danger of it'. On the current status of stem cell research, the surgeon was of the opinion that although, it is not a 'pseudo-science … we are still far away from knowing' the safe amounts of stem cells that can be injected into the body. 'The problem here is we do not know how many cells we need yet' per dose, stated the doctor.[102]

Another surgeon at a private hospital in New Delhi (henceforth 'Hospital A') who has treated children with cerebral palsy using

---

[102] Society of Regenerative Medicine and Tissue Engineering, *Stem 2013*, Proceedings of the Ninth Annual Conference on Biotechnology, Bangalore, last modified 1 February 2013, https://www.srmte.org/index.php/events/stem-2013.

autologous (belonging to the patient) bone marrow stem cells also admitted to the lack of knowledge on the appropriate dosage of stem cells. The 'interesting thing in this procedure', the surgeon stated is that 'till today there is no definite fixed protocol' for treatment and the 'cell count' is unknown. Nonetheless, 'this is what the world is doing', he added. While patients and caregivers must understand that there is no 'standard textbook procedure' for treatment, the doctor always assures them that the procedure is 'pretty safe'. Autologous stem cells are 'your own cells so there's no risk of any reaction or anything going worse', stated the doctor. The stem cells are injected into the spinal cord at Hospital A and like the surgeon at the Bangalore conference, this doctor also describes the procedure as 'simple' because it requires nothing 'more than an injection'. The doctor does not guarantee 'any improvement' with the treatment but he thinks the response has 'definitely' been 'positive'. He prescribes three rounds of stem cell injections at the cost of about two lakh Rupees for the entire treatment. After the first injection the doctor lets the patients or caregivers decide if they want to continue with all three rounds:

> We tell them [patients/caregiver] is that, okay we go through one shot and then you see the result yourself. Like you are happy, I'm happy and if both of us feel like going through more stages and since this is not a drug so you have no limitation … like we get twenty percent, forty percent improvement with this particular schedule, we can always repeat after six months or after a year or so…. (Personal interview on 25 January 2013)

Another clinician in New Delhi has also used autologous bone marrow stem cells to treat children with neurodevelopmental conditions like cerebral palsy. The doctor is the proprietor of a foundation for children with disabilities. The foundation runs a clinic on its premises (henceforth 'Clinic B') that advertises experimental stem cell treatment on its website. The treatment is provided at nursing homes in the city and the clinic is responsible for making appointments and following up with patients after the treatment. The doctor only advises stem cell treatment for his patients when standard therapy 'has failed'. He acknowledged that there is currently 'no cure' from these 'off-label' or 'not officially approved' stem cell treatments and 'it is too early to say' if there will be proven therapies in the near future. 'So it's something like you know a Pandora's box we have opened,' stated the doctor.

Like Hospital A, the doctor at Clinic B, does not guarantee patients successful results, but the treatment so far has shown some promise. The stem cell injections, according to the doctor had improved the cognitive abilities and muscular movement of the children undergoing treatment 'by one grade'. This slight improvement had a positive effect on the interaction between child and parent:

> When a child with cerebral palsy [for example] gets a better understanding the motivation increases, the child also tries to improve and the parents find it easier to communicate with the child and get the child to do what he wants. (Personal interview on 15 January 2013)

The doctor at Clinic B prefers if his clients are educated about stem cell research before agreeing to the treatment. He stated:

> I want only English speaking parents who have gone to the net. They have done all the studies they want. I'll tell them the links to go. It [families/parents] has to be highly educated [background] otherwise we refuse. Highly educated Internet savvy people to whom I'll tell the the websites to visit. They have to read Google scholar. (Personal interview on 15 January 2013)

At another private hospital in Delhi (henceforth 'Hospital C'), several spinal cord injury patients have either tried stem cell therapies elsewhere or have heard about stem cells from the Internet and the media. A spine specialist at this hospital who conducts stem cell trials for spinal cord injuries, stated:

> When they [patients] ask us about the pros and cons [of stem cell treatment] we reply the same as we do to you so that they are not biased in their decision making. Many have heard of stem cell therapy … then there are others who ask us to go for stem cell therapy. So we tell them it's still experimental, and there is no evidence that there is a beneficial effect, however, there is evidence of side effects but still many of them, because we don't do it, go elsewhere to get the procedure done because so many people are claiming on their websites and in advertisements about the beneficial effects. (Personal interview on 24 December 2012)

Nikolas Rose and Carlos Novas have argued that in the context of current developments in biomedicine, individuals are obliged to be fully informed about current scientific research using their own resources and acumen. Additionally they state, governments promoting health education among their citizens to encourage personal responsibility for one's well-being is not new. What is different though in the present, Rose and Novas believe is that people are actively seeking 'biological explanations' of their conditions, or those they care for, through acts of individual choice.[103] The Internet and other media assist in this understanding of hope that 'is not passive', state Rose and Novas, but one in which the process of recovery is also the responsibility of patients and their caregivers. Online direct-to-consumer advertising of stem cell treatments and patient videos and blogs recounting positive stories of experimental stem cell therapies makes the much needed but distant therapeutic breakthroughs in the techno-science seem almost imminent. The political economy of hope in which the key actors of the stem cell industry operate, therefore, demands a willingness to take risks in order to produce desired outcomes.[104]

The spine specialist at Hospital C was critical of the 'undue hype' created by the media on stem cells and its irresponsible reporting on the subject: 'The media listens to claims without proper documentation…. If today I make a claim that I have done this procedure on a patient and he has started walking, the media will publish it … without a thorough investigation and without understanding what harm they can do by doing this' (Personal interview on 24 December 2012).

News stories on the exaggerated successes of stem cell research, are 'not only bad media practice', stated the doctor, but they also 'have a profound effect on the patients, on their psyche, on raising false hopes'. Many of his patients 'are already devastated not only physically but also economically. They have spent a lot on their treatment … many sell their fields or whatever they have left in the hope that they will get cured'. This unfounded hope created around stem cell therapy, prevents patients 'from participating whole heartedly' in treatment that is 'currently available' for spinal cord injuries, stated the doctor. Standard therapy for spinal cord injuries involves surgery and rehabilitation through exercise. In cases of complete spinal cord injury where the neurons show no signs of regeneration, the currently available treatment does not promise recovery. Why would patients 'go through the rehab process', the doctor explained, when they can hope for a miracle cure or the chance to be able to walk again (Personal interview on 24 December 2012).

[103] Rose and Novas, 'Biological Citizenship', 446.
[104] Rose and Novas, 'Biological Citizenship', 452.

The prospect of a scientific breakthrough in stem cell research is very 'exciting' for doctors like the spine specialist who are limited by existing therapies that currently bring little relief to severely injured patients. Yet, the expectation of this techno-science seems premature. We still do not know how stem cells are regulated. We still do not understand the mechanism that regulates the stem cell and how, for example, a stem cell could 'get converted into a neuron and not a scar tissue that results in more pain among patients' (Personal interview on 24 December 2012).

★ ★ ★

State support for India's biotech's future appears to have been an almost predictable outcome within the framework of policies that enabled the privatization of health care, the globalization of the clinical trial industry and the harmonization of intellectual property rules with that of the West. But what is not foreseeable is if all Indians will become equal partners in the future promise offered by emerging biotechnologies such as stem cells. India has about 50 million diabetics, almost one million new cases of cancer every year and 47 million people are believed to be suffering from coronary heart disease.[105] Many among this large unwell population that includes children appear to be willing to expose their bodies to risky stem cell experimentation in the hope of relieving their personal suffering. Several people have died in clinical trials in recent years and there is no way of knowing how stem cells have behaved in the bodies of those who have undergone untested stem cell treatments. If state support for stem cell research would guarantee everyone access to affordable and available clinical outcomes of scientific research the future for patients waiting for cures would seem less bleak than what it is today. The state, thus far, has given no such assurances. People must pay for their health care and those who cannot, will only be provided a package of health services that may not serve their basic health needs. Individuals are left to make their own decisions about medical experimentation through ethical mechanisms such as informed consent. The procedure of consent, intended to assist trial subjects in making informed choices, free of coercion is, however, far from fool proof in the context of India's iniquitous health system. When individuals do consent to participate in a trial, the drugs tested

[105] Dash, *Indian Biotechnology.*

on them may not be marketed in India, leave alone be affordable to most Indians after the trial is over.[106] Clinical trial sponsors give no assurances to trial subjects or the general public that safe and effective drugs tested will be provided to Indians at affordable rates after the clinical trial process is complete.[107]

Even within the market framework that is increasingly the context of health care in India today, the state has done little to ensure that medicines remain affordable to the public. Since India's obligation to TRIPS in 2005, the country has issued only one compulsory license. Compulsory licensing is a provision under TRIPS that gives governments, in the interest of public health, the right to grant generic drug manufacturers the license to produce a product that is still under patent.[108] Today, more than half of India's population does not have access to the medicines they require. To make matters worse, expenditure on drugs account for 81 per cent of private health care spending in rural India and 75 per cent in urban India.[109]

As the Indian state continues to find ways for the nation to be globally recognized for its potential in techno-scientific research and innovation, it gives no indication of a plan for people's access to the benefits of biotechnologies. State promotion of stem cell research is driving the political economy of hope that underlies this highly uncertain techno-science. This hope for therapeutic success appears to be detracting both patient and doctor from what conventional medicine has to offer for currently incurable conditions. 'Even at their most optimistic, current previsions about clinical uses of stem cells are prefaced by the conditional formulas—"It is hoped", it is expected, "it is possible".'[110] Increasingly then, the question that we are confronted with is if Indians will remain experimental subjects or paying patients who are given hope when there may be none.

[106] Sunder Rajan, 'Experimental Values', 67–88.

[107] Nundy and Gulhati, 'A New Colonialism?' 1633–6.

[108] Priya Ranjan, 'The Political Economy of Technoscience'.

[109] Amit Sengupta, 'Universal Health Care in India: Making It Public, Making It a Reality', Occasional Paper 19, Canada, Municipal Services Project, 2013.

[110] Melinda Cooper, 'Resuscitations: Stem Cells and the Crisis of Old Age', *Body & Society* 12, no. 1 (2006): 1–23.

Afterword

Mainstreaming Indigenous Knowledge: Genealogy of a Meta-concept

Dhruv Raina

Conceptual history, according to Koselleck, addresses the development of fundamental concepts underlying and informing a distinctively historical manner of being in the world. In this reflection on history he pointed out that the temporality marking historical processes is multilayered, marked by multiple rhythms of social change that causally determines social reality. This social reality is conditioned by three basic oppositions, the span between birth and death, inner and outer, and above and below. Furthermore, the historicization of conceptual evolution itself reveals the evolution of the language of historians.[1] This shifts the focus from the ontological to the constitutive nature of historical discourse.

In the discussion that follows, I shall look at the different institutional arenas and actors in India, and to an extent outside India, for the current discussion on indigenous knowledge, indigenous knowledge systems, and indigenous sciences, my intention being to explore the mainstreaming of the concept of indigenous science and knowledge.

---

[1] Reinhart Koselleck, *The Practice of Conceptual History: Timing History, Spacing Concepts*, trans. Todd Samuel Presner and others (Stanford University Press, 2002).

In doing so, nevertheless, I do not present a detailed history of these concepts but highlight certain conjunctures in the discussion in India. When seen as a kind of knowledge, indigenous science or knowledge is deployed as a meta-concept, frequently marking the geographical situatedness or social identity of knowledge possessed by a certain community or social group. Furthermore, as someone located within the social studies of science examining social as well as cognitive movements where the boundaries are often quite blurred, there is the academic, critical, and technocratic side to the life-history of any concept. In other words the genealogy of the concept and its subsequent deployment is linked up with constituencies and institutions that include social movements, NGOs, academicians, and international agencies that at times mediate the linkages between the others and the world of politics, and in the process orient its 'mainstreaming'. The neologism 'mainstreaming' would in this context refer to the institutional translation that draws an 'out-group' and its cognitive resources into the epistemological and normative regime of 'globalized science'.

The quest for the indigenous towards the last quarter of the twentieth century was an outcome of several factors, one of them being the disenchantment with modern science and the developmental imagination that had swept across decolonizing nations in the decades after the Second World War. This discourse around development was characterized by a manifold of terms that included science and technology, standard of living, resources, production, poverty, population, planning, market, needs, equality, and environment.[2] One approach towards deconstructing development, it has been pointed out, is to analyse the key words of development discourse, for example, those listed previously, in order to get a better grasp of not just systems of knowledge but cultures characterized as ways of knowing.[3] Indigenous knowledge and indigenous sciences too belong to another manifold or family of terms and concerns which though related to the earlier set,

[2] Wolfgang Sachs, 'Introduction', in *The Development Dictionary: A Guide to Knowledge as Power*, ed. Wolfgang Sachs (London and New Jersey: Zed Books, 1993), 1–5.

[3] Arturo Escobar, 'Development and the Anthropology of Modernity', in *The Postcolonial Science and Technology Reader*, ed. Sandra Harding (Duke University Press, 2011), 269–89.

often in a relation of opposition, foreground the negative outcomes that were seen to be the underside of development.

ENGAGING THE META-CONCEPTS OF INDIGENOUS KNOWLEDGE/SCIENCE

In an elegant counterfactual construction of another science for India, Shiv Visvanathan puts the following words into the mouth of a fictitious Indian scientist disenchanted with the green revolution in India and scientific agriculture: 'Our science should have begun in the forest or in our fields. Instead of the engine, our knowledge should have grown around the seed, the leaf or the cow. We forgot that these were our metaphors and that is why we defoliated the forest.'[4] Discussions on indigenous knowledge systems are replete with metaphors taken from the agricultural or horticultural world to connote not just the multiplicity of resources and their uses, but are equally embedded within multiple life-forms as well. The 'coconut' for one is often used as such a metaphor for 'scientific and techno-logical knowledge developed and used by indigenous and traditional ecological participants worldwide'.[5] The term indigenous knowledge in some national contexts is used to play up a kind of indigenism or projects a radical alterity. But those who creatively play around with the coconut as metaphor of a different knowledge ideal, underscore the fact that its origins are contested, and that its history of migration celebrates its rooting and subsequently naturalization in a multitude of natural landscapes.

What could be said of the diversity of coconut species or peoples could then be extended further to the notion of indigenous knowl-edge and there is no single notion or construct that captures its diversity. Rather it connotes an entity that encompasses the complex-ity and heterogeneity of peoples, tribes, and ethnicities which have

[4] Shiv Visvanathan, *A Carnival for Science: Essays on Science, Technology and Development* (Delhi: Oxford University Press: 1997)

[5] Jennifer D. Adams, 'One Hundred Ways to Use a Coconut', in *Cultural Studies and Environmentalism: The Confluence of EcoJustice, Place-Based Education and Indigenous Knowledge Systems*, ed. Deborah J. Tippins, Michael P. Mueller, Michiel van Eijck, and Jennifer D. Adams (New York: Springer, 2010), 331–6.

been 'subjugated in their ancestral lands to those who are still removed from their land today'.[6] As a result it is sometimes also alluded to as the '... traditional knowledge of indigenous peoples'.[7] This definition too is contested across cultural, regional, and national contexts as well as across academic/non-academic and indigenous/non-indigenous perspectives. This suggests that the valency of the concept is dependent upon cultural context and its application. Nevertheless, environmental scientists, policy makers seek out knowledge of indigenous peoples. Commercial applications premised on an instrumental conception of indigenous knowledge, for example, bioprospecting, involve scientists working in concert with the practitioners of ethno-medicine, herbal biotechnology, and the pharmaceutical sectors. Beyond the instrumental and reified notions is a conception of indigenous knowledge as '... complex systems that arise from indigenous cosmologies and are based on indigenous epistemologies'.[8]

However, despite the polysemy of the term there are several features that give it some unity albeit this characterization is a negative one. We have become quite used to the term non-Western science in the history of sciences as well as its critique. And so indigenity connotes '...ways of knowing that are nonEurocentric [sic], often place-based and often subjugated to Eurocentric cultural and political worldviews'.[9, 10, 11] The notion of place-based or geography is crucial and this is one of the senses in which the term attempts to disengage with Eurocentric theories of knowledge. This embeddedness of theories of knowledge implies a place which is '... an external, physical construct

[6] Jennifer D. Adams, 'One Hundred Ways to Use a Coconut', 332.

[7] Kelly Bannister and Maui Solomon, 'Indigenous Knowledges', in *The Palgrave Dictionary of Transnational History: From the Mid-19th Century to the Present Day*, ed. Akira Iriye and Pierre-Yves Saunier (Basingstoke: Palgrave Macmillan, 2009), 523–6.

[8] Bannister and Solomon, 'Indigenous Knowledges'.

[9] Jennifer D. Adams, 'One Hundred Ways to Use a Coconut', 333.

[10] Sandra Harding, *Is Science Multicultural? Postcolonialisms, Feminisms, and Epistemologies* (Bloomington: Indiana University Press, 1998).

[11] Ladislaus M. Semali and Joe L. Kincheloe, 'What Is Indigenous Knowledge and Why Should We Study It?' in *What Is Indigenous Knowledge: Voices from the Academy*, ed. Ladislaus M. Semali and Joe L. Kincheloe (New York: Falmer Press, 1999), 3–58.

as well as internally constituted'.[12] Thus, in this conception knowledge carries the signature of specific lands and traditions, and the notion of tradition is not to be construed as the antonym of modernity in a social theoretic framework where tradition–modernity constitute an essential dichotomy. Here the notion of tradition connotes the idea of '... being evolved and matured over time and ... as a characteristic or attribute of the peoples concerned'. Secondly, because of its being anchored in space it is '... not seen as limited in time or space but continually evolving and responding to the natural world'.[13] This means that it does share features of scientific knowledge though it is substantially different. Lest the term be conflated with knowledge as rooted in the cultures of 'first peoples' alone, the term neo-indigenous was employed to signify a '... long-standing, noneurocentric mainstream culture'.[14] This gives the notion of indigenous knowledge a provenance that it shares with non-Western and when qualifying forms of knowledge underscores the relationship of subalternity with respect to knowledge from the European world.

Furthermore, arguing against the common trope of so-called traditional knowledge systems whose development had been arrested, which in turn marked their distinction from the sciences, indigenous knowledge systems came to be characterized as '... dynamic: new knowledge is continuously added. Such systems do innovate from within and also will internalize, use and adapt external knowledge to suit the local situation'.[15] For those in social movements indigenous knowledge is '... stored in people's memories and activities and is expressed in stories, songs, folklore, proverbs, devices, myths, cultural values, beliefs, rituals, community laws, local language and taxonomy, agricultural practices, equipment, materials, plant species, and animal breeds'.[16] Thus indigenous knowledge refers to the intergenerational

[12] Adams, 'One Hundred Ways to Use a Coconut', 333.

[13] Bannister and Solomon, 'Indigenous Knowledges', 523.

[14] Glen S. Aikenhead and Masakata Ogawa, 'Indigenous Knowledge and Science Revisited', in *Cultural Studies of Science Education* 2, no. 3 (2007): 539–620.

[15] Louise Grenier, *Working with Indigenous Knowledge: A Guide for Researchers* (Canada: IDRC, 1998), 7.

[16] Grenier, *Working with Indigenous Knowledge*, 7.

accumulation of all those experiences and aspects that define a form of life that is passed on from one generation of indigenous peoples to the next. And further they all share a '... commonly held belief that there is an interdependence and holistic relationship existing between the physical and spiritual worlds'.[17] This could well conjure up a notion of knowledge as scientia sacra. In other words '... integral to these belief systems is that the physical and spiritual well-being of present and future generations is dependent upon maintaining the physical and spiritual health and vitality of the environment in which we live'.[18]

As far as the mode of transmission is concerned indigenous knowledge (hereafter IK) is '... shared and communicated orally, by specific example and through culture. Indigenous forms of communication and organization are vital to local-level decision making procedures and to the preservation, development and spread of IK'.[19] Hence its transmission is endangered because of the breakdown of community and its institutionalized forms of communication, and a new form of life itself has endangered this knowledge form. I shall not go into the causative factors responsible for the loss of IK because these themselves could vary with cultural and regional context. But here again the diversity of causes could be summed up under the rubric of rapid modernization that unleashes in its wake processes of cultural homogenization.[20]

Three transnational trends have externally driven attempts to define and understand indigenous knowledge that 'includes increasing globalization of markets and goods across national borders ... increasing connectivity and global access to improved communication aiding the formation of transnational networks addressing collective interests and rights', and finally the global transition to a 'knowledge-based economy' and escalating multi-stake holder interest in intellectual property protection mechanisms. As a result transnational networks have played an increasingly important role in highlighting the significance of IK, in turn prompted by a perceived sense of endangerment arising from a '... significant increase in the appropriation and commodification of some aspects of indigenous

---

[17] Bannister and Solomon, 'Indigenous Knowledges', 523.
[18] Bannister and Solomon, 'Indigenous Knowledges', 523.
[19] Bannister and Solomon, 'Indigenous Knowledges'.
[20] Grenier, *Working with Indigenous Knowledge*.

knowledge systems, which has catalyzed international calls for protecting indigenous knowledges and associated land resource, cultural
heritage and human rights of indigenous peoples'.[21,22] This then
explains an additional factor that prompted the insertion of the
protection of indigenous knowledge in the Bangalore communiqué
discussed ahead, which paves the way for institutionally mainstreaming 'indigenous science' through efforts orchestrated by transnational
institutions and councils of big science.

FROM THE CRITIQUE OF DEVELOPMENT TO SUSTAINABLE DEVELOPMENT: THE ASCENT OF THE INDIGENOUS

The decades 1950–90 are referred to as the age of development and
the 1990s mark the end of that epoch. The 1950s was the beginning of
the decade of decolonization and the idea of development '... oriented
emerging nations in their journey through post-war history'. Four
decades later the idea of development is another ruin '... in the intellectual landscape', but it still dominates the scenery and '... development
talk still pervades not only official declarations but even the language
of grassroots movements'.[23] However, Sachs in a manner of speaking situates the launching of the age of development on 29th January
1949 when Harry S. Trumann designated the Southern hemisphere as
'underdeveloped areas', a label which provided the cognitive basis for
'... arrogant interventionism from the North and pathetic self-pity in
the South'.[24] The Truman doctrine as a programme of development
was universally embraced and was based on the idea that increased production was the path to posterity. This euphemistic broadside on the
dangers of communism offered a supposedly democratic way out of
poverty and underdevelopment through the '... vigourous application

[21] Bannister and Solomon, 'Indigenous Knowledges', 524.

[22] Manjusha S. Nair, 'Defining Indigeneity: Situating Transnational
Knowledge', *World Society Focus Paper Series* (Zurich: World Society
Foundation, 2006).

[23] Sachs, 'Introduction', 1.

[24] Sachs, 'Introduction', 3.

of modern scientific and technical knowledge'.[25] This imaginary shaped reality and technology '... carried the promise of redeeming the human condition from sweat toil and terrors'. This hope was naturally belied within a decade and half of its launch. Secondly, this idea of development captured the imagination of decolonizing countries in order to' escape the ever present clasp of colonialism, captivated as they were by the promise of '... the economic development of the underdeveloped countries'.[26] Development assigned a historical task to the decolonizing nations namely that of 'catching up' with the industrialized world, Sachs even went so far as to suggest that its hidden agenda was 'nothing else than the Westernization of the world'.[27] This could only be accomplished at the cost of the eclipse of multitude of languages '... the standardization of desires and dreams' accomplished through the universalized instruments of market, state, and science.[28]

If the 1950s to the early 1970s was the golden age of scientism in India; the 1970s was the decade of its undoing, marked as it was not just by a disenchantment with modernity and development but a reckoning with the cultural rootedness of scientific ideas as the notion of value-free science began to disperse into thin air. As one of the neo-Gandhian critiques would write that science '... for us ... was always another culture's product, a recognizable entity. We eventually came to see it as an epoch-specific ethnic (Western) and culture-specific (culturally entombed) project, one that is politically directed, ... invading and distorting and often attempting to take over, the larger more stable canvas of human perceptions and experience'.[29] What Alvarez was forgetting was the century long process involved in domesticating modern science to the Indian environment. Science in this scheme was visualized like colonialism that '... subjects, undermines, subordinates, and then replaces what it eliminates with its own exemplar'.[30] In any case, the key words in academic, policy, and activist networks were alternate development, alternate modernity, and alternate technology–

[25] Escobar, 'Development and the Anthropology of Modernity', 269–70.
[26] Escobar, 'Development and the Anthropology of Modernity'.
[27] Sachs, 'Introduction', 1–5.
[28] Sachs, 'Introduction', 4.
[29] Claude Alvares, 'Science', in *Development Dictionary*, 219–20.
[30] Alvares, 'Science'.

these coruscated within a larger postcolonial critique of Western science and modernity.

Further what should have been a critique of the global commoditization of scientific knowledge became a cultural specific critique of the sciences. The seductive power of science over the decolonizing imagination derived from its intimate linkage with development, which was construed as '... science's latest associate in the exercise of its political hegemony'. The relationship between science and development was traced back to the industrial revolution when both came to be linked in a particular kind of resource mobilization.[31] This mode of resource mobilization generated a crisis through two mistaken identifications. The first had to do with the identification of '... culturally perceived poverty of earth-centred economies with the real material derivation that occurs in market-oriented economies'. The second had to do with the identification of increasing commodity production '... with providing better human sustenance for all'. As a result, development in the post-war world according to Shiva has been driven by the conversion of nature '... into a resource and the use of natural resources for commodity production and capital accumulation'.[32] Consequently, the protection of natural rights of men and nature everywhere would express itself as a revolt against development which in a manner of speaking would also manifest itself as a '... revolt against modern science and the violence it symbolizes'.[33] In fact this became the rallying trope of the critique of westernized development and science in the 1980s and was reflected in the title of a very important collection of essays: *Science, Hegemony and Violence: A Requiem to Modernity.*

In contradistinction, in the 1980s and 1990s the new vision of sustainable development emphasized eco–justice, place-based education and indigenous knowledge all of which were marginalized in national and international environments. This vision was anchored in the 'passion for the Earth and cultural diversity'. The passion itself signified an increasingly felt need for new ways of rethinking the relation between the Earth and its peoples.[34] As Shiv Visvanathan once put it '... ecology

---

[31] Alvares, 'Science', 222.

[32] Vandana Shiva, 'Resources', in *Development Dictionary*, 215.

[33] Alvares, 'Science', 231.

[34] Michael Mueller and Deborah Tippins, 'Prologue', in *Cultural Studies and Environmentalism*, 23.

was a political science and the political goal of ecology was to save agriculture as a way of life'.[35] The ecological sciences have over the last decades, so the argument is formulated, benefitted from these traditional or indigenous knowledge systems that are in their own ways 'collective ways of knowing' and acting or mediating across worldviews that have 'adverse influences and impacts'.[36] In short if through the 1970s and early 1980s development and concern for ecology and the environment were seen as antithetical to each other; sustainable development reconciled this antagonism.

But with the amplification of the developmental crisis throughout the 1980s, IK began to surface in discussions on sustainability and technology transfer. There is very little on IK-based interventions in the 1980s, however by the 1990s it was to become a ground for fruitful research and by the end of the 1990s the time was more or less ripe for preparing the text book or teaching manual where the paradigm would be stabilized for subsequent reproduction. Since a great deal of this work was policy and intervention based, developmental agencies such as IDRC, UNDP took upon themselves the task for paving the way. The textbook or the guide is evidently an important indicator of the notion of IK as prevalent both within and outside the academy. Thus one such characterization of IK suggests that it '... refers to the unique, traditional, local knowledge existing within and developed around the specific conditions of women and men indigenous to a particular geographical area'.[37] The importance of this knowledge arises in the context of the endangered lives of the concerned peoples but also for a larger global community. 'The development of IK systems, covering all aspects of life, including management of the natural environment, has been a matter of survival to the peoples who generated these systems. Such knowledge systems are cumulative, representing generations of experiences, careful observations and trial-and-error experiments.'[38] These knowledge systems are epistemically characterized as cumulative in nature and empirical in practice.

The interest in IK '...emerged in tandem with the politicization of indigenous groups and indigenous-rights movements', wherein the

[35] Visvanathan, *Carnival for Science*, 203.
[36] Visvanathan, *Carnival for Science*, 24.
[37] Grenier, *Working with Indigenous Knowledge*, 6.
[38] Grenier, *Working with Indigenous Knowledge*, 6.

indigenous people began to demand the right to be heard in development-related decisions concerning themselves. As a result IK came to be lauded as '... an alternative collective wisdom relevant to a variety of matters at a time when existing norms, values and laws are increasingly called into question'.[39] It serves as a cognitive resource where development planning has not delivered on promised results, and in our time this goal is sustainable development.[40] The principle of sustenance is itself based on the economy of nature conferring on human societies the material basis of survival through a variety of provisioning mechanisms.[41] As a result sustainability requires that markets and production processes be reshaped by nature's logic of returns, '... not the logic of profits, capital accumulation, and returns of investment'. In other words development 'must be guided by the limits nature imposes on the economy'.[42] One of the many failings of development has been that it has created dependencies on an outside world that 'orders and demands ... but does not truly correspond to development'.[43] Against the backdrop of encroachment that endangers the local, the centering of the indigenous and indigenous knowledge acquires both cognitive salience and ideological legitimacy. The point is further substantiated in the construction of 'traditional technologies' which are made out to be '... effective, inexpensive, *locally available, culturally appropriate*' (emphasis added).[44] Indigenous Knowledge is seen to be able to significantly contribute to strategies for local sustainable development since it is seen to be responsive to local circumstances, experience, and 'wisdom'.[45]

However, this argument sustains development itself and not nature unless of course sustaining development recognizes the limits of

[39] Berks (1993: 7), quoted in Grenier, *Working with Indigenous Knowledge*.

[40] According to the World Commission on Environment and Development (1987: 43), sustainable development is that '... development that meets the needs of the present without compromising the ability of future generations to meet their own demands'. See Grenier, *Working with Indigenous Knowledge*, 12, for the nine objectives of sustainable development.

[41] Shiva, 'Resources', 215.

[42] Shiva, 'Resources', 216–17.

[43] Grenier, *Working with Indigenous Knowledge*.

[44] Grenier, *Working with Indigenous Knowledge*.

[45] Grenier, *Working with Indigenous Knowledge*, 12

nature and being guided by them. In fact, Shiva argues that sustainability originally referred to nature's ability for supporting life which meant recognizing '… the integrity of nature's processes, cycles and rhythms', and the current crisis was a product of the neglect of natural processes and impairing nature's capacity to recuperate. Shiva writes: 'In a finite, ecologically interconnected and entropy-bound world, nature's limits need to be respected; they cannot be set by the whims and conveniences of capital and market forces, no matter how clever the technologies summoned to their aid.'[46] Clearly Shiva is already gesturing to other forms of knowledge that respect nature's limits and economy rather than the economy of the market.

In June–July 1999 UNESCO planned a World Conference on 'Science for the Twenty First Century: A New Commitment' and requested member countries to send in their recommendations for Conference. In India's scientific capital–Bangalore–the National Institute of Advanced Studies, organized an 'International Symposium on Science and Society: A New Social Contract' in January 1999. The meeting involving scientists and policy makers to frame recommendations for the UNESCO meeting to be held at Budapest later in June–July that year. The suggested recommendations came to be referred to as the *Bangalore Communiqué on Science and Society*.[47] The communiqué referred to many aspects of science such as its human dimension, as a knowledge system and then went on to raise issues of equity, access, empowerment, and cooperation. Finally the communiqué concluded with a discussion on the challenges and tasks for the next millennium. However, below are some of the recommendations of the communiqué:

> Concerns for intellectual property rights should not be allowed to exploit the various forms of indigenous and civililzational knowledge so faithfully preserved by indigenous populations over millennia, care being taken to ensure that benefits derived from such biological and intellectual resources are not misappropriated.[48]

[46] Grenier, *Working with Indigenous Knowledge*, 217

[47] 'Bangalore Communiqué on Science and Society', *International Symposium on Science and Society: A New Social Contract*, Bangalore, 23–29 January 1999, www.unesco.org/science/wcs/meetings/apa_bangalore_99. html (accessed 16 November 2015).

[48] 'Bangalore Communiqué on Science and Society'.

The framework of action presented at Budapest included the following recommendations:

1. Establish an international initiative for the conservation and promotion of indigenous and civilizational knowledge systems to enable the Member States to recognize, protect and promote such knowledge systems.

2. Design a series of projects that would recognize the range of influence and economic value of technologies either residing in indigenous and civilizational knowledge systems or emerging from grass-root innovations and study how best to protect and reward such intellectual property leading to a revised international convention on protection of intellectual property rights with special emphasis on indigenous and civilizational knowledge systems.[49]

I do not intend to get into issues of intellectual property rights here, but would restrict the discussion to the interesting internal differentiation the communiqué evokes between two kinds of knowledge. The emphases in the recommendations have to do with the protection of *civilizational and indigenous knowledge*. This, in itself is a reflection of the impress of different social movements and constituencies speaking in the interests of a variety of so-called indigenous knowledge systems, as well as the institutional conflicts these knowledge commitments entail and the negotiation over what is considered civilizational knowledge (that I shall hereafter allude to as IK1) and indigenous knowledge (hereafter IK2). The recommendations of the communiqué were adopted at the World Conference on Science in Budapest in June 1999.

DISCIPLINARY ENGAGEMENTS WITH IK1 AND IK2 IN INDIA

If the 1970s was the decade of the disenchantment with modernity and science in India, it was also the decade in historiography when percolation models of modernization, developmental processes, and circulation of knowledge began to be challenged. As a result questions began to be asked about the agency of the colonized and the rejection of osmotic models of the transmission of knowledge.[50] As

---

[49] 'Bangalore Communiqué on Science and Society'.

[50] Dhruv Raina, *Images and Contexts: The Historiography of Science and Modernity in India* (Delhi: Oxford University Press, 2003).

a result in the history of sciences reception studies took a new turn in favour of rejecting the rational west–irrational east dichotomy and indigenous responses became an important area of study, while anthropologists discussed the epistemological violence colonial systems of education had inflicted upon indigenous ways of knowing.[51] However, since the 1970s the term IK in India served as rubric for speaking of the high traditions of knowledge before colonialism as well as the knowledge systems of tribal populations distributed across the country. These constellations of knowledge can then be mapped on to IK1 and IK2 respectively. In Hess' classification IK1 would correspond to the knowledge of literate, Old World societies and IK2 to that of the New World societies and indigenous Old World societies. The former (IK1) would roughly conform to the high knowledge traditions of Babylon, Greece, Rome, Egypt, ancient India, and China—a world that was abuzz '… with transmissions that followed routes of trade and conquest'. As a result scholars have debated whether these should be considered variations of a single system or local knowledge in historical dialogue over millennia.[52]

The relationship between IK1 and IK2 in South Asia has never been free of conflict with IK1 appropriating the latter as settled society moved into the tribal regions of the country. But the defendants of IK1 emphasize the relationship between IK1 and modern scientific knowledge and the associated politics of domination that has institutionally and epistemically de-priviliged the former. Around this constructed experience of marginalization and erasure an attempt has been underway both within the academy and in the political realm to access a view of these indigenous knowledge systems that are undistorted by the experience of colonialism. In that sense the identity of the Indian nation and its systems of knowledge are traced back to the Vedic age which is the fount and source of Indian nationhood and its defining epistemological regimes. Thus, in a recent encyclopedia on Non-Western Science published by Springer the entry on 'Knowledge Systems in India' points out: 'Traditionally all knowledge systems in India have been traced to the Vedas…. The Vedas are considered to

[51] Bernard S. Cohn, *Colonialism and its Forms of Knowledge: The British in India* (Delhi: Oxford University Press, 1997).

[52] David J. Hess, *Science and Technology in a Multicultural World: The Cultural Politics of Arts and Artifacts* (New York: Columbia University Press, 1995), 186.

be divine revelation.... In Indian knowledge systems, it is the science of linguistics that occupied the central place which, in the West was occupied by mathematics.'[53] The definition of knowledge system here is grounded in that of formal knowledge and the Vedas are centred to be the source of the subsequent evolution of the system much in analogy to the conception of Greece in relation to Western science. In a comparative context this entry sits amid entries on the knowledge systems of the Incas, the Australian aborigines, indigenous populations of the Americas, the Olmec, and so on. Furthermore, this is followed by a brief abstract of the Budapest Declaration on Science and the Use of Scientific Knowledge.[54] It is not as if indigenous people were lacking on the Indian subcontinent, but IK1 appears to have captured the entire space of indigenous knowledge and edged out IK2.

Furthermore, both nationally and internationally the entire debate on indigenous knowledge is pivoted around several registers—the one in the domain of NGOs, social movements, and international organizations whose genealogy is traceable to the humanism of the post League of Nations organizations (UNESCO, UNDP, UNCTAD, and so on). In addition, to which there is also a significant academic take on the subject. At the academic level the cross-cultural study of knowledge was an important trend in anthropology in the 1960s and 1970s, but was not so important within what came to be designated as Science and Technology Studies (STS) till the 1990s, energized at the conjuncture of several disciplinary trends within the social studies of science, anthropology, feminism, post-colonialism, geography, and environmentalism.[55] This moment differed from the earlier one that took Western rationality as the benchmark for assessing other knowledge forms and systems. In the earlier studies traditional knowledge systems were portrayed as utilitarian, value laden, closed, pragmatic, indexical, content dependent, and thereby lacking science's authority, restricted as this knowledge was by the 'social and cultural circumstances of its

[53] A.V. Balasubramanian, 'Knowledge Systems in India', in *Encyclopaedia of the History of Science, Technology and Medicine in Non-Western Cultures*, ed. Helaine Selin (Massachusetts: Springer, 2008), 1182–3.

[54] Helaine Selin, ed., *Encyclopaedia of the History of Science, Technology and Medicine in Non-Western Cultures*, 1191.

[55] Harding, *Is Science Multicultural?*

production. The divide of knowledge systems was the divide between societies'.[56] On the other hand, philosophers such as Stephen Stich and others have been speaking of folk epistemology that sees knowledge systems as products of cultural variation; an idea that earlier was looked upon with contempt by philosophers, until even Western philosophical theories began to be seen as instantiations of folk epistemology.[57]

THE INSTITUTIONAL MAINSTREAMING OF IKS

Consequently, the literature dealing with 'Other Ways of Knowing and Doing' is very vast today and there are a multitude of reasons for engaging with these 'other ways'. Within the new economy of science, and the ideological commitment to sustainable development two reasons acquire salience. The first has to do with the recognition of the limitations of contemporary institutionalized science; which is why it is suggested that such an engagement would make it '... easier to see the possibilities and limitations of the ways in which the human mind organizes the world ... (and) may raise new problems and suggest new methods and topics of inquiry'. If the first recognizes the limitations of science the second is a more pragmatic consideration that seeks to protect other knowledge systems from a more dominating one: '... understanding the knowledge systems of other cultures may help the members of those cultures to resist colonization, either in the ideological or in the political or economic domain.' In other words in order to resist the ideologies of pro-development groups the study of non-Western natural knowledge may help build a better case '... against development projects' that endangers irreplaceable natural resources such as tropical forests.[58]

In the debate on the new production of knowledge Helga Nowotny, Peter Scott, and Michael Gibbons pointed out a transition from a long

---

[56] Helen Watson-Verran and David Turnbull, 'Science and Other Indigenous Knowledge Systems' in *Handbook of Science and Technology Studies*, ed. Sheila Jasanoff, Gerald E. Markle, James C. Petersen, and Trevor Pinch (New Delhi: Sage Publications, 1994), 115–16.

[57] Steve Fuller, *The Knowledge Book: Key Concepts in Philosophy, Science and Culture* (Stocksfield, UK: Acumen Publishing, 2007), 53.

[58] Hess, *Science and Technology in a Multicultural World*, 185.

period of the scientization of society to that of the socialization of science.[59] In several disciplinary academic communities we are already witness to engaging more with histories of knowledge than histories of sciences. But when it comes to the indigenous and ethno, the labels of ethno-mathematics and indigenous sciences still reflect the political struggles within pedagogy in most parts of the South America, Oceania, and Africa.[60] On the other hand, the pragmatics of dealing with problems of a global nature have forced the scientific community to recognize some of the interesting work going on in Africa and elsewhere. The web based journal SciDev Net was established precisely for this task. The network was established by some members of the staff of the leading journal of the sciences *Nature* following in the footsteps of Bangalore Communiqué and the recommendations of the Budapest conference. The project was initially supported by DFID as a project SciDev.net.[61] Later regional networks mushroomed in sub-Saharan Africa, Latin America and Carribean, South Asia, and now more or less embrace the globe. A recent search of the articles, features, and news appearing in SciDev featured 537 items for the key word 'indigenous' that was then used to qualify knowledge, sciences, languages, and people. Similarly the key word 'traditional' threw up 916 items used in turn to qualify medicine, ethno-pharmacology, knowledge, and so on. This reflects the possible mainstreaming of IK and indigenous science has begun in earnest as the recognition in the scientific community dawns that the problems of poverty and sustainability are far too complex and risky to be handled by what is received as big science today.

[59] Helga Nowotny, Peter Scott, and Michael Gibbons, *Re-thinking Science, Knowledge and the Public in the Age of Uncertainty* (Cambridge: Polity Press, 2001).

[60] Ubiratan D'Ambrosio, 'Ethnomathematics and its Place in History and Pedagogy of Mathematics', *For the Learning of Mathematics* 5, no. 1 (1985): 44–8.

[61] 'Science in Development Network (SciDev)', http://www.scidev.net/en/ (accessed 16 November 2015); www.r4d.dfid.gov.uk/Project/3937/Default.aspx (accessed 16 November 2015).

Notes on Editors and Contributors

EDITORS

Sarah Hodges is Associate Professor in the Department of History at the University of Warwick, UK. She works on the social and cultural history of modern South Asia, specifically the politics of health in colonial and postcolonial India (particularly the Tamil-speaking south). Her publications include *Contraception, Colonialism and Commerce: Birth Control in South India, 1920–1940* (2008) and an edited volume, *Reproductive Health in India: History, Politics, Controversies* (2006). She is currently finishing a monograph on medical garbage in India.

Mohan Rao is Professor at the Centre of Social Medicine and Community Health (CSMCH), School of Social Sciences, Jawaharlal Nehru University (JNU), New Delhi. A medical doctor specialized in public health, he has written extensively on health and population policy, and on the history and politics of health and family planning. He is the author of *From Population Control to Reproductive Health: Malthusian Arithmetic* (2004) and has edited *Disinvesting in Health: The World Bank's Health Prescriptions* (1999) and *The Unheard Scream: Reproductive Health and Women's Lives in India* (2004). He has edited, with Sarah Sexton of Cornerhouse, UK, the volume *Markets and Malthus: Population, Gender and Health in Neoliberal Times* (2010).

CONTRIBUTORS

David Arnold is Emeritus Professor of History at the University of Warwick, UK, having previously been Professor of South Asian History at SOAS in London from 1988 to 2006. A founder member of the Subaltern Studies group of scholars, he has also taught at the University of Dar-es-Salaam in Tanzania and at the University of Lancaster, been a visiting professor in Chicago and Zurich, and is a Fellow of the British Academy. His published work has addressed various aspects of social and political history of India, with particular reference to medicine, science, and technology. These include *Colonizing the Body: State Medicine and Epidemic Disease in Nineteenth-Century India* (1993), *Science, Technology and Medicine in Colonial India* (2000); *Gandhi* (2001); *The Tropics and the Traveling Gaze: India, Landscape, and Science, 1800–1856* (2006); and *Everyday Technology: Machines and the Making of India's Modernity* (2013). He also edited *Imperial Medicine and Indigenous Societies* (1988) and *Warm Climates and Western Medicine: The Emergence of Tropical Medicine, 1500–1900* (1996). His current research is on poisons and pollution in India, c. 1830–1984.

Rama Baru is Professor at the CSMCH, JNU, New Delhi. Her research focus is on health policy, international health, privatization of health services, and inequalities in health. She is the author of *Private Health Care in India: Social Characteristics and Trends* (1998) and has more recently edited the volume *School Health Services in India: The Social and Economic Contexts* (2009). In addition, she has published extensively in journals and several edited volumes. She was awarded the Balzan Fellowship by the University College London and the Shastri Indo-Canadian Institute Fellowship. Professor Baru is the regional editor for South Asia for the journal *Global Social Policy*. She has served as a member of research committees for the Government of India, the World Health Organization, the Indian Council for Medical Research, and the Ministry of Health and Family Welfare, Government of India.

Ramila Bisht is Associate Professor at the CSMCH, JNU, New Delhi. She holds an MA in psychology and an MPhil and a PhD in social sciences in health from JNU. Before joining CSMCH in 2008, she has taught at the Tata Institute of Social Sciences, Mumbai. Her research interests comprise issues related to health disparities, women's health; health policy and administration; comparative health policy and health-care reform. She has been involved in several health services research projects and programme evaluation studies. She is the author of *Environmental Health in Garhwal Himalayas* (2002). She was awarded the British Society of Population Studies LEDC Visitor Award, 2010. She was an embedded fellow in 2010 in the ESRC Rising Powers Network Award entitled 'India's Challenge in a Globalizing Healthcare Economy: Social Science Directions'.

Rohini Kandhari is a PhD candidate at the CSMCH, JNU, New Delhi. Her PhD thesis, titled 'Stem Cell Research and Experimentation in India: Mapping Practice and Policy', is a qualitative research study that investigates India's engagement with stem cell technologies through the narratives of the scientist, the doctor, the patient/caregiver, and the policy maker. She has also done her MPhil at CSMCH. Her MPhil thesis was titled 'The Role of Institutional Ethics Committees in Clinical Trials in India: A Study of Selected Hospitals in New Delhi'.

Lakshmi Kutty is a PhD candidate at the CSMCH, JNU, New Delhi. Her research traces the trajectory of the question of nutrition in tuberculosis policy and public health debates. She is the author of 'The "Intractable" Patient: Managing Context, Illness, Health Care', in *Towards a Critical Medical Practice: Reflections on the Dilemmas of Medical Culture Today* (ed. Anand Zachariah, R. Srivatsan, and Susie Tharu [2010]).

Dhruv Raina is Professor at JNU, New Delhi. He studied physics at the Indian Institute of Technology, Mumbai, and received his PhD in the philosophy of science from the University of Gothenburg, Sweden. His research has focused upon the politics and cultures of scientific knowledge in South Asia. He has co-edited *Situating the History of Science: Dialogues* (with Joseph Needham, 1999), *Social History of Sciences in Colonial India* (with S. Irfan Habib, 2007), and *Science between Europe and Asia* (with S. Irfan Habib, 2010). *Images and Contexts: The Historiography of Science and Modernity* (2003) was a collection of papers contextualizing science and its modernity in India. He is also the co-author of *Domesticating Modern Science* (with S. Irfan Habib, 2004). His most recent book is *Needham's Indian Network* (2015). He has been a Fellow of the Wissenschaftskolleg zu Berlin, a Visiting Fellow at the Max-Planck-Institut für Wissenschaftsgeschichte (MPIWG), Berlin, and the first incumbent of the Heinrich Zimmer Chair for Indian Philosophy and Intellectual History at Heidelberg University. In addition, he has been a Visiting Professor for several years at the Maison des sciences de l'Homme and Université de Paris Diderot, Paris. He has also been a visiting faculty at the ETH, Zurich, and Guest Fellow at Caius and Gonville College and Centre for Research in the Arts, Humanities and Social Sciences, Cambridge University, UK.

Priya Ranjan has recently completed his doctor thesis, 'The Political Economy of Technoscience: A Study of Medical Biotechnology in India', at the Centre of Social Medicine and Community Health (CSMCH), School of Social Sciences (SSS), JNU, New Delhi. He is currently working on clinical trials in India.

Altaf Virani is a doctoral candidate at the Lee Kuan Yew School of Public Policy, National University of Singapore. He was previously a researcher at the Indian Institute of Management, Bangalore. His research interests focus on health systems and policy, particularly themes of knowledge-translation, public management reform, and organizational productivity and performance.

Rebecca Williams is Lecturer in Medical History at the University of Exeter. Her research interests are in the history of medicine and development in modern South Asia, particularly the history of population control in India. She received her PhD from the University of Warwick, UK. Her doctoral thesis looked at the establishment of India as a 'laboratory' for population control during the 1950s, focusing on a well-known family planning project known as the Khanna Study. She has also written about family planning during the 'Emergency' of 1975–7 in India; this work has been published in the *Journal of Asian Studies*.

Index